GRASSROOTS ENVIRONMENTAL ACTION

People's participation in sustainable development

Edited by
*Dharam Ghai and
Jessica M. Vivian*

London and New York

First published 1992
Paperback edition first published 1995
by Routledge
11 New Fetter Lane, London EC4P 4EE

Simultaneously published in the USA and Canada
by Routledge
29 West 35th Street, New York, NY 10001

© 1992, 1995 United Nations Research Institute for Social Development

Typeset in Linotron Garamond by
J&L Composition Ltd, Filey, North Yorkshire
Printed and bound in Great Britain by
Mackays of Chatham PLC, Chatham, Kent

British Library Cataloguing in Publication Data
A catalogue record for this title is available from the British Library.

Library of Congress Cataloging in Publication Data
Grassroots environmental action: people's participation in
sustainable development / edited by Dharam Ghai and Jessica M.
Vivian.
p.   cm.
Includes bibliographical references and index.
1. Environmental policy—Developing countries—Citizen
participation—Case studies.   2. Economic development—Environmental
aspects—Citizen participation—Case studies.   I. Ghai, Dharam P.
II. Vivian, Jessica M.
HC59.72.E5G73   1992
363.7′058′091724—dc20
91–41512

ISBN 0–415–07762–1
0–415–12703–3 (pbk)

CONTENTS

CONTENTS

vi

FIGURES

TABLES

CONTRIBUTORS

Jayanta Bandyopadhyay is an ecologist presently at ICIMOD (the International Centre for Integrated Mountain Development) in Kathmandu. He received a Ph.D. in Physical Metallurgy from the Indian Institute of Technology in Kanpur in 1974, and has worked extensively in the area of natural resource conflicts and environment-development relationships, especially in the context of the mountains. He has numerous research publications on issues related to water resources, forests, people's environmental action and ecological research.

Jutta Blauert received her Ph.D. in Rural Sociology and Environmental Studies from Wye College, University of London. Her research interests include sustainable agriculture and indigenous and peasants' movements, especially in Mexico and the Andean countries. She is currently Honorary Research Fellow at the Institute of Latin American Studies at the University of London.

Antonio Carlos S. Diegues is a social scientist who is presently Research Director of the Programme on Research and Management of Wetlands in Brazil, which is supported by the University of São Paulo, IUCN and the Ford Foundation. Previous work with the United Nations and other international agencies has brought him to numerous countries in Africa, Asia and Latin America.

Philippe Egger holds a Ph.D. in Political Science (1980) from the University of Grenoble. He worked successively with UNICEF in Niger on drought rehabilitation programmes and with UNRISD researching people's participation before joining the International Labour Organization in Geneva in 1983. There he has been working on participatory organizations, poverty alleviation and rural employment policies and programmes.

Dharam Ghai has been Director of the United Nations Research Institute for Social Development (UNRISD) since 1987. He was previously Chief of Research at the ILO and Director of the Institute of Development Studies at the University of Nairobi. He studied at Oxford and received his Ph.D. in Economics from Yale, and is the author of several books and numerous articles on a wide range of development issues, particularly poverty and rural development.

Marta Guidi is a social anthropologist with several years' experience researching development issues in Latin America. Her interests include environmental issues, indigenous communities, women and migration in Latin America. She spent three years working with the Proyecto Mixtepec.

John Kurien is an activist-researcher who has organized fish-workers in Kerala State and also undertaken research into the techno-economic, socio-ecological and political dynamics of marine fisheries in India and Asia. He is a founding member of the International Collective in Support of Fishworkers. He has been a member of various task forces on fisheries and natural resources in Kerala, and is currently Associate Fellow at the Centre for Development Studies in Trivandrum, India.

Charles Lane has spent ten years as a development worker in Tanzania. His research on Barabaig land tenure was part of his doctoral studies at the Institute of Development Studies, University of Sussex. He subsequently campaigned with the Barabaig for their land rights. He is presently Senior Research Associate with the International Institute for Environment and Development (IIED) in London.

Jean Majeres holds a Doctorate in Development Economics and Sociology from the University of Paris. After two years with the European Development Fund in Brussels, he joined the International Labour Organization in 1974, first in Zaire as an adviser on human resources development planning, and since 1977 at its headquarters in Geneva where he has been working on rural employment policy and development issues.

Julio Moguel is an economist and Professor at the Universidad Nacional Autónoma de México. He has written extensively on urban and rural collective action over the past ten years.

Michael Redclift is presently Reader in Rural Sociology at Wye College, University of London, and Research Coordinator of the ESRC programme on Global Environmental Change. He has published several books and numerous articles on environment and development issues, especially in Latin America.

Charles A. Reilly is Vice President of the Inter-American Foundation. He has written on issues of development, public policy, taught at the University of California, San Diego, and directed development NGOs in Central America. Among the publications he has authored or co-authored are *Religión y Política en México*, *In Partnership with People*, and *Development, Public Policy and Local Politics in Brazil*.

Michael Ståhl has been involved in development research in Africa for many years. He has been a Lecturer of Political Science in Ethiopia, a Senior Research Fellow with BRALUP in Tanzania, the Director of the Scandanavian Institute of African Studies and Research Officer with the Swedish Agency for Research Cooperation with Developing Countries (SAREC). He is presently head of the Regional Soil Conservation Unit of the Swedish Development Cooperation Office, and is based in Nairobi.

Enrique Velázquez, an economist, has published on political and ecological issues in Mexico, and has been coordinator of the Comité de Defensa y Preservación Ecológica de Durango (CDyPE) since 1987.

Jessica M. Vivian is a Research Fellow at the United Nations Research Institute for Social Development (UNRISD), where she has been responsible for coordinating the research programme on Sustainable Development through People's Participation in Resource Management since 1989. She undertook doctoral studies in the Department of City and Regional Planning at Cornell University, and previously spent several years as a development worker in the Republic of Kiribati and in the Sudan.

Ruth Ammerman Yabes is an Assistant Professor at the Department of Planning of Arizona State University at Tempe. She received her Ph.D. in City and Regional Planning from Cornell University. She spent several years in the Philippines, as a development worker and subsequently conducting research on irrigation societies.

FOREWORD

Maurice F. Strong
Secretary-General
United Nations Conference on Environment and
Development

Experience increasingly shows that the imperative transition to sustainable development cannot be made without the full support of the community and the participation of ordinary people at the local level. The global issues of environment and development demand international responses, intergovernmental cooperation and regional efforts in areas such as capacity building. But most actions, even on such major global risks as climate change, must be taken at the local and national levels.

The systemic nature of environmental issues and the globalization of the economy require that even actions which have their primary impacts at the local level must be taken in a global context. Just as global environmental problems arise as the result of the accumulation of individual transgressions, so too local-level actions such as resource management are the very foundations of successful sustainable development policy. At the national level, a government may set a regulatory framework and economic and fiscal policies which provide positive incentives for environmentally sound and sustainable behaviour, but, in the final analysis, it is the activities of individuals that count.

Economic change is the key to sustainable development, and environmental costs must be internalized. Equally important is the integration of social issues into the policy-making process. A growing number of governments and institutions require environmental impact assessments to be carried out on prospective projects, and, increasingly, the best practice is to ensure that the social effects are taken into full account. Policymakers must be provided with information that will inform them of the true costs and implications of their intended actions. The participation of people at the local level in meeting the challenge of sustainable development has the great merit of providing a mechanism for taking into consideration local conditions and social issues at every stage of the planning process. As the case studies in this volume demonstrate, development plans that do not have the support of those affected rarely succeed.

This book highlights the benefits to be gained by drawing on the vast experience of traditional resource management systems and sounds a warning of the risks of failure for projects that do not adequately reflect local value systems and opinions. It comes as a timely contribution to the literature on a subject that is increasingly acknowledged as being of vital importance.

Among the recent developments that demonstrate an understanding of the significance of people's participation in resource management is a recent decision in the UNCED preparatory process. The Preparatory Committee of the United Nations General Assembly, responsible for the preparations and negotiations leading up to the Conference on Environment and Development in June 1992, has recognized the importance of the contributions made by indigenous peoples and local communities to the UNCED process, and more broadly to sustainable management of resources. It will draw on the traditional knowledge and resource management practices of indigenous people and local communities as contributions to environmentally sound and sustainable development, and intends to integrate them into the programmes expected to result from this first 'Earth Summit'.

In the final analysis it is at the grassroots level where the 'action' is and upon which sustainable development depends. We have a lot to learn from skills that have been developed and practised at the local level. This book is a valuable guide, providing sound analysis and documenting a variety of relevant practical experience.

Plymouth University
The Charles Seale-Hayne Library

Customer ID: *****27428

Title: Grassroots environmental
action : people's participation in
sustainable development /
ID: 9002467700
Due: 16/01/2015 23:59:00 GMT

Total items: 1
05 Dec 2014
Checked out: 8
Overdue: 0
Hold requests: 0

Due date of borrowed items subject
to change. Please check your
University email regularly for recall
notices to avoid incurring a fine.

ACKNOWLEDGEMENTS

Many people have been involved in the process which has led to the publication of this book, and we are grateful for the advice and support they have contributed. Elizabeth Cecelski formulated the original research programme outline which launched the body of research work at UNRISD on sustainable development and participation. Research plans were further developed at a workshop in Geneva in May 1990, and special thanks are due to Shimwaayi Muntemba, Michael Ståhl and Michael Redclift for the contributions they made both to the workshop and to the formulation of the research framework.

Several people read and commented on parts of the manuscript, including Solon Barraclough, Cynthia Hewitt de Alcántara and Peter Utting. Fred Buttel and John Friedmann each read the entire manuscript, and made extensive and extremely useful comments. We are grateful for all the comments received, which helped us to make substantial improvements to the volume. However, we take full responsibility for the deficiencies which remain.

Finally, we would like to acknowledge the considerable production assistance we have received in various stages of the work on this book. Monica Bradley, Adrienne Cruz, Françoise Jaffré, Irene Ruiz and Josephine Yates all provided assistance which not only expedited the production of the manuscript, but also improved the final copy. Rhonda Gibbes merits special mention for her detailed proofreading and checking of references of the manuscript, as does Wendy Salvo, who assisted substantially with the organization of the research planning meeting and the administration of the subsequent research programme.

Dharam Ghai
Jessica M. Vivian

1

INTRODUCTION

Dharam Ghai
and Jessica M. Vivian

PURPOSE AND OVERVIEW

This volume deals with the threat of environmental degradation in the Third World, and with the wide range of actions people take on the local level to manage and protect their natural resources – activities which have the potential to help reverse, arrest or prevent environmental decline. When such actions are undertaken they stem primarily from a concern with livelihood: with people's desire to maintain or improve levels of living which, in much of the Third World, depend to a large extent on the ability to make productive use of natural resources. The book is thus concerned with what has come to be known as 'sustainable development'.

It is the goal of this collection to contribute to a better understanding of how people participate in the management of their environment, both in situations where they are motivated, encouraged and supported by outside actors such as the state or international organizations, and in situations where they must formulate their plans and conduct their activities in spite of the neglect, resistance or even active opposition of external forces. The main premise underlying the selection of this focus is that sustainable environmental management, and, where necessary, environmental rehabilitation, can only occur where active local-level support exists. It is clear that managing resources sustainably on the local level is essential for achieving the global goal of sustainable development: although more macro-level activities are also important, it is the combined impact of the small-scale activities – either constructive or destructive – undertaken by vast numbers of individuals which will determine the fate of many resources and ecosystems, especially in the Third World.

The volume emphasizes the diversity of ways in which people act to influence the utilization of their resource base, and thus employs a broader definition of the term 'participation' than is current in some circles. Much early use of this term discussed participation primarily as an alternative, more efficient management style to be created and promoted by agencies wishing to increase the success rates of their projects. As the case studies presented here illustrate, however, people participate in resource management in a variety of ways which do not always involve outside intervention. The development of rules and structures by a society to ensure that resources are not overexploited by any individuals or groups, for instance, is a kind of participation which forms the basis of many communities. In addition, people often organize to oppose the resource management priorities of external agents, rather than to cooperate with them. Such social mobilization is an especially active and visible form of participation in resource management practices, as well as a particularly important one.

This collection does not claim to establish that such local-level participation in resource management – which may be called, more specifically, 'grassroots environmental action' – is sufficient by itself to prevent or reverse environmental degradation in situations where national policies or global-level ecological changes create major destructive forces. But local-level participation is necessary for sustainable development. In certain cases, this is obvious: in lands needing reforestation, for instance, the sheer labour requirements of such an undertaking necessitate wide-scale cooperation with tree-planting projects. Grassroots participation, however, is also required in virtually every situation where environmental degradation threatens – from situations in which pastoralists must maintain social controls to prevent the overexploitation of resources, to those in which industrial pollution will only be constrained by activities which are initiated at the community level.

A second premise which has influenced the approach taken in this volume is that, although the importance of people's participation for sustainable development has recently become increasingly acknowledged, there is as yet insufficient analysis of the multiple dimensions that such participation involves. Specifically, the emphasis in much of the literature on sustainable development and participation has not been on people's initiatives as keys to finding *solutions* for environmental problems; rather it has been on people's *lack* of participation (for instance in forestry schemes) as an *obstacle* to the solution (by outsiders) of environmental problems. In other words, there has been much more attention paid to ways in which local people can be persuaded to provide the necessary labour input into environmental projects designed outside the community than to ways in which grassroots initiatives, stemming from indigenous environmental concerns, can inform the development of more successful projects – or can supplement or even replace the project approach. Of course, local people cannot and do not always provide solutions to their environmental problems, but the argument made by this collection is that the potential of grassroots environmental action has been unduly neglected, and that a wider recognition of the ways in which people act to protect their environments can only advance the search for sustainable development.

This argument is set forth more fully in the first section of the volume. The chapter by Redclift discusses the importance of developing a framework for analysing the issues involved in sustainable development which takes into account the need for people's participation and the utility of local-level environmental management. He argues that sustainable development should be seen as an alternative to the prevailing development paradigms, rather than a modification of them. Redclift's approach calls for a substantial rethinking of the theoretical basis of sustainable development, which, he asserts, should include not only economic, but also political and epistemological dimensions. Thus questions of participation, including empowerment and local knowledge systems, must be addressed by any programme concerned with environmental issues in the context of development.

The chapter by Vivian builds on Redclift's arguments, demonstrating in more specific terms the importance of traditional resource management systems and locally based popular environmental initiatives in situations of actual or potential environmental degradation. She demonstrates that, in some cases, such an approach leads to conclusions which contest the universal validity of models of environment and development relationships which have been widely accepted – including the 'tragedy of the commons' scenario, the assumed simple causal relationship between population growth and environmental degradation, and the link between poverty and environmental stress. Thus a more accurate picture of the ways in

3

which grassroots action affects the environment is essential for the formulation of effective policies for sustainable development.

The remaining three sections of the volume are devoted to case studies which provide concrete illustrations of the role of grassroots environmental action in resource management and sustainable development. Three types of local-level activity have been distinguished, providing the structure for the collection. The in-depth studies of traditional systems of resource management in Part II are followed, in Part III, by analyses of social action centred around environmental issues, while the last section examines lessons regarding local participation in resource management or conservation projects.

TRADITIONAL SYSTEMS OF RESOURCE MANAGEMENT

The case studies in Part II come from Tanzania, the Philippines and Brazil. They document the mechanisms and structures of traditional resource management systems, as well as people's reactions to pressures put on them, in three very different ecosystems. Lane's chapter documents the ecological knowledge of the pastoralists of the Hanang plains of north-eastern Tanzania, and provides evidence of the efficiency of pastoralism in the area. He argues that, in the Barabaig case, the use to which pastoralists put their land is preferable, in social, environmental and economic terms, to the capital-intensive agricultural scheme which threatens to displace them in favour of large-scale monocropping.

Yabes looks at one example of the traditional irrigation societies which have supplied water for much of Asian agriculture for centuries, and are based on communal, participatory management systems. She demonstrates the environmental and productive damage which results when such traditional management arrangements are supplanted by 'modern' irrigation infrastructure in the name of 'development', ignoring local experience in resource management and overturning social structures developed through generations to enforce rules for sustainable resource use. In the case study discussed here, local non-cooperation and active opposition to the irrigation project led to a revision of the project plans and the creation of a more participatory approach to project management.

Diegues's study documents the traditional resource management

4

practices of two Brazilian wetlands communities, and explores the ecological and social impacts of the disruption of these practices by state-supported schemes for increased levels of resource exploitation. This chapter shows the environmental impacts of large-scale agricultural schemes for the wetlands ecosystem of the Marituba floodplain, and the resulting adverse effects of the projects on local livelihoods. A similar pattern is shown by events in the Guaporé valley, where people making a living from the sustainable extraction of rainforest products are being displaced by colonists and cattle ranchers. In both cases, local people have organized to protect their livelihood, and have been able to develop initiatives which would protect both their livelihood and the integrity of their environment.

In spite of the very different ecosystems, social systems and locations of these three studies, they all tell a similar story: one in which traditionally sustainable resource management systems have been the object of outsiders' plans for 'improved' production systems developed without consultation with the local people. The results were also similar in the three cases: increased environmental degradation, in two of the cases combined with the virtual exclusion of the traditional resource managers from sharing in the benefits of the new projects, led to protests and political action which attempted to alleviate the environmental damage done and improve the prospects for sustainable resource use in the region. Such action was undertaken because conservation practices are a part of the social structure of these communities, which have a keen awareness of the environmental constraints with which they have lived for generations.

Taken together, these three studies provide empirical evidence that not only can traditional resource management practices remain sustainable, but also that they are commonly more efficient, in their long-term productive capacity, than is generally recognized. Thus, although the studies on their own do not provide a sufficient basis for making generalizations about the productivity of all traditional resource management systems, they do provide sufficient evidence to justify a re-examination of the long-term trade-offs between traditional systems and capital-intensive development schemes which make no attempt to benefit from the environmental expertise of local communities. Even if the short-term gains from such projects, in terms of increased productivity, are clear, their long-term results are much more uncertain than is commonly acknowledged. Obviously, this is not meant to imply that all traditional practices are sustainable or even desirable, that development efforts are unnecessary, or that traditional societies always have the means and ability to achieve local-level sustainable development. Rather, these studies make the point that a better understanding of – and communication with – local communities can help to overcome some of the most common environmental and social problems associated with efforts to promote development.

SOCIAL ACTION AND THE ENVIRONMENT

Part III in many respects further substantiates the points made in Part II, although the analyses presented here focus in more depth on the dynamics of the struggles waged by marginalized groups for control and preservation of natural resources when the environment is threatened by outsiders. The case studies come from Mexico and India – two countries with a particularly strong tradition of grassroots activism.

The two chapters from Mexico describe both urban and rural movements which have had environmental impacts. Moguel and Velázquez analyse what began as an urban-based revolutionary movement which had, as one of its aims, the acquisition of new settlements for its members through invasions and occupation of land. The movement's specifically environmental concerns originated with the necessity of confronting the polluters of the community water supply, and the success of this initial environmental activity was followed by the establishment of a wider campaign to raise environmental consciousness.

The chapter on two communities in rural Oaxaca by Blauert and Guidi describes movements which grew out of efforts by indigenous communities to defend their livelihood, assert control over their natural resources and struggle for cultural autonomy. Under the leadership of teachers and returned migrants, a series of small-scale activities relating to soil conservation, water harvesting, reforestation, land tenure, health, education and training was initiated in the Mixteca. The second study concerns the UCIZONI, a movement of indigenous communities in the Mixe region, which was a successor to a series of organizations which sprang up in the 1960s and the 1970s to struggle against the loss of natural resources and declining living standards triggered by modernization processes. It built up alliances with sympathetic groups in the state administration, the environmental lobby and the academic community in its fight for restoration of land and forest reserves to indigenous groups, protection of human rights and preservation of local culture.

The two chapters from India again use examples from different ecosystems to demonstrate that grassroots action in defence of the environment can have a significant impact, even in the face of the opposition of entrenched interest groups whose profit lies in the overexploitation of resources. The chapter by Kurien discusses the struggle for livelihood waged by the fishing community in Kerala in south India in response to the introduction of modern trawlers, which depleted stocks and led to declining incomes for the fisherfolk. With the help of outside social workers and church groups, the fishing community responded to the crisis by creating a trade-union type organization to exert pressure on the government to change its fisheries policies and to undertake collective action to improve traditional technology and better manage marine resources.

The chapter by Bandyopadhyay seeks to illuminate the dynamics of the process of widening ecological consciousness among the forest people through a study of two celebrated environmental movements in the Himalayan region. The driving force behind the Chipko movement was the threat posed to local livelihood by vanishing forest resources, but people's determination to resist commercial and state interests in logging was reinforced by their growing consciousness of the wider deleterious effects of deforestation in the mountainous areas. In this the local communities were strengthened by support from scientific research and the increasingly powerful domestic and international constituency for environmental conservation. The movement against the construction of a high dam in Tehri arose in the same area, and again, the impetus behind the movement was the threat to the livelihood of the people who would be forced to relocate. But as in the Chipko case, evidence provided by the findings of scientific research regarding the destructive potential of such a dam was a powerful trigger in mobilizing grassroots action and external support.

The four chapters in this section, taken together, illustrate the conflict generated by the competition over control and use of natural resources between local communities and state and capitalist interests. This conflict results from the inequity, social injustice and environmental degradation which too often results from the conventional approaches to development. In Kerala, trawler owners, backed by state subsidies, deplete the marine resources and thus jeopardize the livelihood of the traditional fishing community. In the Himalaya, logging contractors and energy companies, with state support, threaten the life-support systems of the forest dwellers. The indigenous communities in Mexico have over the decades been deprived of their access to natural resources by the penetration of capitalist interests in logging, livestock and commercial agriculture assisted by state subsidies, colonization programmes and provision of infrastructure such as roads and dams. Such destructive practices are allowed to persist in part because those who benefit from them do not have to bear their costs, and this fact explains the necessity, in so many circumstances, of grassroots action for environmental preservation. The deprivation and destruction of resources compel indigenous communities to organize and offer resistance to the processes of modernization, and the studies presented in this section are unanimous in demonstrating that such resistance is not in defence of 'environment' in the abstract, but is rather inspired by people's need to safeguard their livelihood and, in a broader sense, their way of life.

LESSONS FROM ENVIRONMENTAL PROJECTS

The chapters in Part IV focus on environmental projects initiated by the state or grassroots support organizations, generally with funding from external sources. While the case studies in Parts II and III emphasize the themes of conflict, resistance and political struggle, and analyse environmental action initiated at the community level, those in Part IV stress the practical aspects of the preparation and implementation of projects designed to ameliorate living standards through improvements to the environment. As these chapters demonstrate, while in some circumstances local communities struggle to maintain the integrity of their environment in the face of destructive pressures emanating from outside the community, in other circumstances state and development agencies are themselves concerned with improving the sustainability of resource management on the local level, while local communities have neither the necessary resources nor the option to improve their environment without external support. Grassroots participation, however, remains the most important element determining the outcome of sustainable development projects.

The chapter by Ståhl discusses the ambitious attempt made by the previous Ethiopian government to arrest and reverse the severe environmental degradation of the highland areas caused by rising human and livestock population, increasing fuelwood demand from urban areas and inappropriate policies, and exacerbated by the extreme poverty of the local population. Financed by the World Food Programme and some bilateral donors, the programme was the second largest of its kind in the world, but, despite some undoubted successes, it failed to make a significant dent in the vast environmental problems of the highlands. At the heart of this failure was the inability of the state to establish a genuine partnership with the peasantry in the formulation and implementation of the programme. The projects were conceived and executed in a highly centralist and bureaucratic manner, and, although they perceived the need for soil and water conservation, the peasants were unwilling to participate in the schemes without the provision of food. State policies of maximum resource extraction, primarily for its wars against insurgencies, the lack of secure land tenure and the sheer poverty and exhaustion of the peasantry all contributed to local people's feeling that participation in the project would not be to their benefit, and thus limited the impact of the project.

The last two chapters in the book draw on a wide range of experiences to provide more general lessons on participation in community-level sustainable development projects which are supported and/or initiated by external agents. Egger and Majeres report on a number of environmental projects assisted or studied by the International Labour Office, and covering a wide range of activities such as soil conservation, irrigation, reforestation, establishment of orchards and vegetable gardens, protection of marine resources and construction of wells and minor roads. They are typically labour-intensive, rely on local materials and tools, and are initiated either by the government or by grassroots organizations. In either case there is some external financial or technical assistance. A key feature of the government-initiated projects is the active participation of the communities affected in the planning and implementation of agreed upon activities. In other cases, the central role is played by grassroots support organizations such as the Six-S in Burkina Faso, Process in the Philippines and Bhoomi Sena in India.

The chapter by Reilly contains some reflections stimulated by the experience of environmental and other development projects in

Latin America supported by the Inter-American Foundation (IAF). The IAF provides financial and technical assistance to non-governmental organizations and local groups for participatory development initiatives, and has supported projects in such areas as agroecology, social forestry, marine resource conservation, water-shed management, waste recycling, energy conservation, appropriate technology, environmental education and legislation. The author illustrates the characteristics of these projects, and highlights the critical role played by 'meso' organizations which mediate between national and local-level groups. Such grassroots support organizations provide technical and professional counselling, serve as mechanisms for delivery of credit and social services and help establish contacts with sympathetic and influential individuals and institutions at the national and international levels. These experiences further demonstrate the importance of building multiclass alliances in the pursuit of environmental objectives, and point to the critical role played by technical knowledge, participatory research and field experimentation in understanding and finding solutions to environmental problems.

The three chapters which form this section broaden our understanding of the scope and causes of environmental degradation. While the earlier chapters emphasized the rural and natural resource dimensions of environmental problems, the material contained in Part IV extends their scope to urban sanitation, water supply and waste disposal and recycling. Furthermore, environmental deterioration is shown to result not only from exogenous pressures, but also from an interaction between increasing human and animal populations, limited natural resources and outmoded production systems.

AN ALTERNATIVE PERSPECTIVE ON ENVIRONMENTAL DEGRADATION

If the case studies described above include three different areas of focus, the essays in this volume are brought together by their common concern with the relationship between people and their environment. Collectively, they offer the basis for the formulation of what might be considered an alternative model of the processes which are involved in environmental degradation. The remainder of this chapter will describe this alternative model, contrasting it with what we consider to be the prevailing explanations for current environmental problems in the Third World, and will attempt to draw out the policy-relevant implications which are suggested in the process.

Perceptions of how local people interact with the environment have evolved substantially over recent years. Earlier writing on this subject tended to imply that people were generally negligent in terms of their resource management practices, or else were uninformed about the importance of the environment and the consequences of their actions for their resource base. This view has given way to another stylized scenario which now commonly appears in the literature and, less explicitly, in policy documents which address environmental concerns: one in which poverty, population pressures and external forces drive people to undertake environmentally destructive practices. Thus, although care is taken not to 'blame' poor people for environmental degradation (the current metaphor is that the poor should not be held accountable for environmental problems any more than the foot soldier should be blamed for the ravages of war), local communities are often portrayed as reactive, buffeted by forces beyond their control, and helplessly watching as their own activities contribute to the destruction of their chances for a better life in the future.

The case studies in this book, however, suggest that an alternative model might be equally as valid as the one outlined above. These essays describe situations in which people are knowledgeable, concerned and active about their resource management options and decisions – in short, situations in which people in local communities are actors who participate vigorously in the determination of their own future and that of their environment. The book thus suggests an alternative scenario regarding the process of environmental degradation, which will be outlined below. Admittedly this scenario is also stylized (as, indeed, all models must be), and we must stress here that we do not regard it as *the* model of the process of environmental degradation – rather, it may be considered as an alternative perspective, one which, under some circumstances, may be more helpful in understanding the dynamics involved in the generation and progression of environmental problems, and may contribute more successfully to policies which will alleviate them.

The scenario we will suggest has four components. First, the case studies in this book demonstrate the importance of the environment for vast numbers of people in the rural areas of the Third World countries. Environmental resources are valued as a source of livelihood by groups as diverse as the fisherfolk of Kerala, the forest dwellers of the Himalayas and the Amazon, the herders of Tanzania, the nomads of the Sahel and the peasants of the Philippines and Mexico. Furthermore, the social significance of the environment is seen clearly in the relationship which traditional communities have developed with their natural resources. Not only are cultivation, herding, fishing, hunting and gathering direct sources of livelihood, but these activities also furnish materials for building, medicine and manufactures, and thus natural resources possess cultural and spiritual significance for many communities.

Second, because of the economic, social and spiritual importance of the environment, physical and social systems of resource management have been developed throughout the world. Where such resource management systems have been successful, they maintain a delicate balance between the livelihood needs of the people and the integrity of the environment. To achieve this balance, local communities have developed complex and ingenious systems of institutions and rules regulating the ownership and use of natural resources. Local knowledge, skills and technologies built up over time and handed down from one generation to another have ensured the continued functioning of these systems of resource management.

Third, the historical processes of foreign conquest, colonialism and settlement and associated 'modernizing forces' have undermined or modified the operation of such resource management systems in much of the world. Confiscation of resources by foreign invaders, assertion of control over common resources by the new authorities, establishment of plantations, ranches and logging concessions, introduction of new products and technologies, commercialization of agriculture, and construction of mines, dams, irrigation canals and roads and railways have occurred in a wide variety of situations. Such developments have often had a devastating impact on the livelihood of local communities, traditional systems of resource management and the quality of the environment. Livelihood has been jeopardized through deprivation and degradation of natural resources. Indigenous management systems have suffered from loss of control over resources by the local communities and the establishment of new property regimes by the central authorities. The environment has been degraded by excessive exploitation of resources, topographical changes brought about by infrastructure projects, intensive use of land and water, cultivation and grazing in fragile terrain and pollution caused by new technologies and inputs. These processes have been further intensified by unprecedented increases in human and animal populations.

The fourth aspect of this model is the resistance offered to these processes by the adversely affected social groups in their struggle to safeguard their livelihood. In some cases, people have taken to direct action such as invasion and occupation of land. More often, local communities have sought to bring pressure on the state and on competing commercial and industrial interests to restore ownership or use of resources to their members, or to desist from further exploitation or degradation of contested resources. For this they have had to rely on collective action through indigenous or newly created organizations. They have also attempted to create broad coalitions to press their claims, enlisting support from workers' and peasants' associations, social activists, religious bodies, environmental groups and the scientific community. Increasingly they have turned to the donor community and foreign ecological and human rights groups for support in their struggle for social justice and environmental conservation. Pressure on the state and competing economic interests have been further intensified by a variety of methods such as non-cooperation, strikes, demonstrations and protest marches, participation in the electoral process, alliances with supportive political groups, use of media and reliance on experimentation and scientific research.

In terms of concrete achievements, the record of this resistance is a mixed one. In some cases the people's claims have been recognized and their goals largely met. This is so, for instance, with the struggles waged by the indigenous groups in Durango for control of pollution and initiation of an environmental rehabilitation programme. Likewise the Chipko movement was successful in securing suspension of commercial logging for a number of years. The farmers in the Philippines succeeded in restructuring the modern irrigation project to safeguard their interests and reinforce the traditional system.

In some other cases, there have been partial victories. The fisherfolk of Kerala succeeded for a time in imposing a monsoon ban on trawler fishing and in securing financial and technical support from the state agencies. In the Marituba floodplain, indigenous groups have succeeded in slowing the march of modernization while the effects on the environment of planned investments are being assessed. Similarly, the rubber tappers in Guaporé region have been successful in halting deforestation and have proposed the creation of extractive reserves managed and controlled by the local people.

In some cases the resistance put up by the local communities has been to no avail. The Barabaig have failed to win back their land or to persuade the authorities to agree to their other demands. Despite sustained resistance over decades, the Garwahl people threatened by the construction of the Tehri dam have not succeeded in having the decision rescinded, although their protests may have contributed to delays in completion of the project.

THE IMPLICATIONS FOR POLICY APPROACHES

Again, we must emphasize that the stylized description of the social processes associated with environmental degradation given above is not intended to account for environmental conditions in all situations. Instead, we are arguing that it should be recognized that this scenario does occur, because these processes have considerably different implications for policy approaches than does the prevailing model of the dynamics of environmental degradation. In short, different models of a problem suggest different solutions, and although the analysis presented here does not yield specific, detailed policy conclusions, nor does it provide the 'key' to sustainable development, it does point out several broad issues associated with policy formulation which should be considered by grassroots organizations, local and national authorities, and international agencies.

First, this book argues that people's participation is essential for sustainable development: increasingly, local livelihood requirements necessitate the rehabilitation and conservation of the resources available to local communities, and people's ability and will to undertake such efforts on the local level are prerequisites for successful environmental management. At the same time, because of the wider repercussions of environmental degradation, national authorities, key social groups and the international community as a whole also have a vested interest in protecting and improving natural resources. Thus the interests of local communities in secure livelihoods converge with those of others in environmental improvement. It is therefore both important and timely to consider the ideas suggested by the studies in this volume for more participatory and sustainable patterns of development.

Second, these essays provide a contrast between two alternative approaches to development. The predominant approach pursued in developing countries has been characterized by excessive centralization, large-scale investment and modern technology, and has often resulted in sharp inequalities and widespread impoverishment. It has frequently been environmentally destructive and socially disruptive, with unregulated industry and concessions to capitalist interests contributing both to environmental degradation and the dispossession and impoverishment of indigenous people. The alternative approach to development, which is exemplified by the grassroots environmental movements described here, is characterized by small-scale activities, improved technology, local control of resources, widespread economic and social participation and environmental conservation.

Third, these chapters clearly demonstrate the inappropriateness of reliance on purely technical and economic 'fixes' as means of achieving development goals. Several of the studies bring out the environmentally destructive impact of large-scale infrastructure projects and agricultural investments made by capitalist and state enterprises. It is becoming increasingly clear that many such projects do not realize the benefits, even in purely economic terms, which had been predicted at the planning stage, because of the declining and even negative long-term rates of return which result from the environmental degradation they cause. The long-term unsustainability of large-scale projects means continued inputs of capital are necessary, particularly to cover unforeseen pressures of environmental cycles. At the same time, the displacement of traditional communities which such projects often require create intra- and inter-community conflict, and put new stresses on social structures. There is therefore a need, both on the part of the state and of funding agencies, to re-examine modern development initiatives which reduce local control over resources in the interests of increased productivity. More thorough cost-benefit analyses, which better take into account both environmental factors and the social costs which result from the disruption of traditional systems, are called for. Often, extensive consultation and genuine collaboration with local communities can reduce the environmental and social costs of large-scale projects. More generally, it is important to develop policies and procedures to ensure that those who benefit from such projects bear the full costs of their economic activities, and do not pass them on to other sections of the society.

Fourth, while many large-scale projects have been promoted on the grounds that traditional systems of resource management are economically inefficient and environmentally destructive, the case studies presented here provide further substantiation of the findings of recent research on the equity, efficiency and sustainability of many indigenous systems of resource management. Successful management systems have necessarily maintained an equilibrium between resource extraction and resource preservation which is designed to minimize the risk that environmental 'capital' will have to be expended. Traditional resource management techniques may therefore seem inefficient to outsiders in a given year, but typically a system which allows for excess capacity in times of plenty is necessary to guard against resource overexploitation under the more difficult environmental conditions which periodically occur. The negative effects attributed to indigenous resource management systems are in fact often associated with their breakdown – due in large part to a loss of autonomy and of control over resources suffered by local communities – with the resulting desperate search for livelihood under new and impoverished circumstances further reinforcing a spiral of resource degradation.

The studies thus indicate a need to re-evaluate the productivity of traditional management systems, and to better account for the costs of establishing alternative forms of livelihood to people affected by their disruption. These essays suggest, further, that an appropriate response to situations in which local-level resource management practices have become unsustainable might be to examine ways of restoring and re-invigorating previously effective resource management systems, rather than seeking to replace them with systems which are based on foreign models, and which involve the creation of new and untested technical and social structures.

Fifth, several studies demonstrate that people are not always in a position either to recreate previously sustainable resource management practices which are no longer viable, or to develop, organize and promote alternatives to development policies and programmes. As migration, displacement and increased integration into other cultures has resulted in the loss of ecological knowledge, and as changing production systems or social structures or increased population pressures has made some traditional resource management systems unviable or irrelevant, the importance of introducing new management techniques into areas affected by environmental degradation has increased. Environmental rehabilitation and conservation activities, however, are investments – they require the foregoing of present consumption to build up future productive capacity – and many of the people affected by the process of environmental degradation do not have the surplus resources necessary to undertake such activities. This is perhaps the strongest argument for national and international programmes of assistance for environmental projects, whether in the form of food, materials, tools or cash payments for work. It has proven rather difficult, however, for external agents to successfully introduce appropriate techniques and practices. The type and structure of the organization used to promote and support such techniques have been demonstrated to be keys to their effectiveness. Of particular importance is the establishment of lines of communication with the local level which will enable external agents to correctly evaluate the impact of the proposed changes.

CONCLUSION

This book argues that people's legitimate interest in the conservation of their resource base must be recognized and supported – not only because this is their basic right, but also because it is a pragmatic course to take in the interests of achieving sustainable development. We are not attempting to romanticize indigenous societies, nor to deny the existence of environmentally destructive community-level activities in some circumstances. However, we are arguing that it is essential to recognize the critical importance of indigenous organizations of marginalized people for environmentally sound development under a wide variety of conditions. Such organizations can function both as vehicles of resistance to environmentally destructive practices and as agencies for the promotion of alternative, sustainable approaches to social and economic development, and they are vital for the success of initiatives for environmental preservation, rehabilitation and amelioration. The roles they may play in different situations include working for local empowerment, mobilizing resources, providing mechanisms for the delivery of inputs and services to members, education, training, experimentation and developing participatory research approaches.

Certain conditions are generally conducive to the successful articulation of local environmental concerns through grassroots environmental movements, although these vary according to social and political contexts. Of cardinal importance is the unity, strength and perseverance of organizations of marginalized groups, as well as the development and maintenance of participatory structures within such organizations. The support of sympathetic individuals and institutions has been critical in most cases, with pressure from international sources becoming an increasingly important factor in the outcome of struggles over resources. Likewise, the studies bring out the role played by outside sympathizers in the launching of grassroots environmental activities, and demonstrate the importance, for the successful pursuit of environmental and livelihood objectives, of the forging of broad-based alliances with other supportive groups such as workers' and peasants' associations, social activists, religious bodies, national and international environmental societies, the scientific community, political parties and the media.

Appropriate outside assistance can reinforce the capabilities and effectiveness of grassroots environmental organizations. Such assistance may be based on a recognition of customary rights of ownership and use of resources, while in other situations there may be a need to elaborate new frameworks of rules and regulations governing the ownership and management of resources to ensure equity, efficiency and sustainability. In many circumstances, it is necessary to seek ways to raise labour and resource productivity. An effective way of doing this would be to build development efforts more explicitly on a base of local knowledge, skills and technology, and further participatory research and experimentation in marrying local and modern knowledge will have a particularly important role to play in adapting traditional systems in ways which will promote secure and satisfactory livelihood opportunities, as well as a healthy and sustainable environment.

Perhaps the most important conclusion to emerge from these studies, however, concerns the complex and varied role of the state in the environmental equation. The existence of a democratic space allowing the expression and defence of community rights and claims has proven to be a crucial factor influencing successful grassroots environmental action. A multiparty system appears to offer increased opportunities for social groups to influence state policies, and the presence or absence of democratic space also helps determine, in different degrees, the strategy and tactics adopted by indigenous organizations. These may include the establishment of alliances with existing parties, direct participation in the electoral process, mobilization of support from other social groups, the use of media to publicize environmental causes, and the organization of demonstrations, protest marches and strikes. The essence of these activities is to persuade or pressure the state to intervene on behalf of the communities through adoption of new legislation, to withdraw support for capitalist enterprises, to provide financial resources and technical and social services to local communities or to terminate infrastructure projects or colonization programmes.

Thus the studies presented here indicate that sustainable development requires that local communities enjoy genuine autonomy, have control over adequate resources, and, in some cases, that they be provided with financial and technical assistance to restore their resource base and re-establish their control over resources. The studies in this volume provide several illustrations of the creativity and innovative capability of indigenous management systems. They demonstrate that sustainable development requires people working together and communicating with one another at the local, national and international levels. Finally, they provide further argument, if any is needed, for the importance of promoting and supporting democratic and equitable social and political systems in the context of development.

Part I

APPROACHES AND CONCEPTS

2

SUSTAINABLE DEVELOPMENT AND POPULAR PARTICIPATION: A FRAMEWORK FOR ANALYSIS

Michael Redclift

INTRODUCTION

Both 'sustainable development' and 'environmental management' have become buzz-words in development policy circles, but the discussion surrounding these terms pays scant attention to the way in which people in developing countries participate in the management of their resource base and, through their participation, help to transform the practice of environmental management. This chapter, in addressing these issues, seeks to correct two kinds of bias which exist in much of the sustainable development debate. First, there is a bias towards 'managerialism' rather than resource management, stemming from a top-down approach to local-level development. Second, there is a tendency to treat 'sustainable development' as merely a variation of the prevailing Northern, economic-centred world view of development problems, and to see sustainability as a goal which can be attained through making adjustments to the standard development models.

This chapter, in contrast, will argue that the concept of sustainable development needs to be recognized as an alternative to the prevailing view, rather than a modification of it. The approach taken here reflects a way of examining resource conflicts – through political economy – that some might not share. The emphasis is placed on the structural determinants of local-level decision-making, at the local, national and international levels, rather than on a more 'human resources' or interactional approach. At the same

23

time, the analysis emphasizes that what distinguishes environmental concerns in the North from those of poor areas of the South is not simply material conditions, but different epistemologies, different systems of knowledge.

The first sections of this chapter analyse the concept of sustainable development, and seek to enlarge the conceptual discussion on this topic in order to take more account of some of the inconsistencies and limitations of the definitions currently available. The current thinking in environmental economics, which has gained favour within some international development agencies, and which emphasizes the use of calculations of the environment's value, is critically discussed. The economists' rather limited technical treatment is compared with a more thoroughgoing account of the economic, political and epistemological dimensions of sustainable development.

In this context, some of the new approaches which outside development agencies are currently taking towards local-level environmental management are briefly discussed. Next, the chapter examines some instances of conflicts over resource use which have prompted popular participation and struggles to gain greater local control over the environment. The analysis focuses on situations in which natural resources are highly valued and have been heavily contested politically.

The final section of the chapter outlines an approach to contested environments which departs radically from the analysis of most development agencies by focusing attention on power and political mediation in the resolution of environmental conflicts at the local level. In this section the chapter tries to incorporate some experiences of poor people's participation in resource management in order to set out a framework for analysis that takes into account both the need for popular participation and the utility of local-level environmental management as complementary facets of the problem. It is hoped that, through addressing the political problems associated with local resource management, as well as through developing a more rigorous analysis of the terms under which poor people and their environments are incorporated within development policy, we will begin to identify the potential for determining better policy interventions which is contained in the struggles and resistance of the rural majorities in the South.

SUSTAINABLE DEVELOPMENT:
CONCEPTS AND CONTRADICTIONS

The problem with using the term 'sustainable development' is that it has proven difficult to formulate a definition of it which is comprehensive but not tautological, and which retains analytical precision. In this it is similar to many terms in the development lexicon, whose very appeal, it can be said, lies in their vagueness. 'Sustainable development' means different things to ecologists, environmental planners, economists and environmental activists, although the term is often used as if consensus exists concerning its desirability. Like 'motherhood' and 'God', sustainable development is invoked by different groups of people in support of various projects and goals, both abstract and concrete.

One of the sources of the conceptual confusion surrounding the term 'sustainable development' is that no agreement exists regarding what exactly is to be sustained. The goal of 'sustainability' sometimes refers to the resource base itself, and sometimes to the livelihoods which are derived from it. Some writers refer to sustaining levels of production, while others emphasize sustaining levels of consumption (Redclift 1987). This divergence in emphasis is important since what makes continued 'development' unsustainable at the global level is the pattern of consumption in the rich countries, while most policies designed to tackle development problems, including those which fit within the 'sustainable development' idiom, are essentially production-oriented.

The different uses made of the concept of sustainable development reflect varying disciplinary biases, distinctive paradigms and ideological disputes. In our view there are also at least two sets of contradictions which soon become evident when sustainable development is discussed.

First, embedded in much of the 'sustainability' thinking is an important difference of emphasis. For some writers, the principal problem to be addressed is that 'human progress' carries implications for nature itself, and should cause us to re-examine the 'ends' of development, as well as the means (Devall and Sessions 1985). Others view sustainability as a serious issue because nature is a major constraint on further human progress. They are concerned, basically, with the constraints that will be imposed on the conventional growth model if the warnings we receive from the environment, the 'biospheric imperatives', are ignored. The solution,

25

according to this view, is either to develop technologies which avoid the most dire environmental consequences of economic growth, or to take measures to assess and 'price' environmental losses in a more realistic way, thus reducing the danger that they will be overlooked by policymakers.

Second, when 'sustainable development' is considered within a North–South framework, attention must be paid to the contradictions imposed by the structural inequalities of the global system (Brundtland Commission 1987; Redclift 1987). Green concerns in the North, such as alternatives to work and ways of making work more rewarding, can often be inverted in the South, where the environment is contested not because it is valued for its amenities or aesthetic value, but primarily because its exploitation creates economic value.

In the North, natural resources are also a source of value, and conflict between those who want to exploit them for commercial gain and those who wish to conserve the 'countryside' is often highly charged. However, the very fact that conservation issues are given increasing weight in planning decisions in the developed countries bears witness to the shift in priorities which occurs in the course of 'development'. In urbanized, industrial societies, relatively few people's livelihoods are threatened by conservation measures. The 'quality of life' considerations which play such a large part in dictating the political priorities of developed countries surface precisely because of the success of industrial capitalism in delivering relatively high standards of living for the majority (but by no means all) of the population.

In the South, on the other hand, struggles over the environment are usually about basic needs, cultural identity and strategies of survival, rather than about providing a safety valve within an increasingly congested urban space. Under these circumstances, when the individual and household are forced to behave 'selfishly' in their struggle to survive, there is no point in appealing to idealism or altruism to protect the environment.

SUSTAINABLE DEVELOPMENT ALTERNATIVES

Of the two major trends in sustainable development thinking, one, exemplified by the economic approach taken by Pearce et al. in *Blueprint for a Green Economy* (1989), fails to take into consideration the contradictions discussed above. 'Sustainable development',

has been said to be

in this view, is treated as a modification of traditional development strategy, rather than an alternative to it, and this approach is therefore limited in scope and application. The second major trend, exemplified by the Brundtland report, *Our Common Future* (Brundtland Commission 1987), treats sustainable development as alternative concept of development, and therefore, in the end, shows more promise.

A common point of departure for a discussion of sustainable development is to define it as what Barbier (1989) refers to as *sustainable economic development*. This is an optimal level of inter-action between three systems – the biological, the economic and the social – which is achieved 'through a dynamic and adaptive process of trade-offs' (Barbier 1989: 185). Many economists, notably David Pearce, also emphasize the *trade-offs* between systems, or between present and future needs, as the key issue (Pearce *et al.* 1989). In similar terms it is argued that 'sustainable economic development involves maximizing the net benefits of economic development, subject to maintaining the services and quality of natural resources over time' (Pearce *et al.* 1989), and that '[sustainable development] is development that maintains a particular level of income by conserving the sources of that income: the stock of produced and natural capital' (Bartelmus 1987: 12). For economists interested in the environment, then, procedures such as environmental accounting, which aim to give a numerical value to the environment and to environmental losses, are essential instruments for the achievement of greater sustainability.

In Chapter 3 of *Blueprint for a Green Economy* Pearce and his colleagues argue, from a declared interest in environmental quality, that environmental improvements are equivalent to economic improvements 'if [they] increase social satisfaction or welfare' (p. 52). The resolve of these economists is to demonstrate that there are economic costs in ignoring the environment. This approach is growing in influence within international development agencies such as the World Bank, the United Nations agencies and the Overseas Development Administration (ODA) (see World Bank 1987, 1988a, 1988b). Although all of these organizations have been strongly criticized in the past for funding development projects with very damaging ecological effects, such as cattle ranching in Central America, their new approach has, in a relatively short space of time, become almost synonymous with effective environmental management in many people's estimation.

One of the main problems with this view of environmental management is that it works better for developed than for developing countries. Most neo-classical economists use the 'willingness to pay' principle as a means of assessing environmental costs and benefits, and Pearce argues that the emphasis in environmental policy should be shifted towards this principle to avoid future damage to the environment (Pearce *et al.* 1989: 55). It is not hard to appreciate some of the difficulties in applying the new environmental economics when we consider developing countries. As Pearce *et al.* (1989) demonstrate, there is widespread popular concern about the environment in the North, where environmental quality is often placed before economic growth in surveys of public opinion. In the South, on the other hand, immediate problems of acquiring subsistence needs preclude extensive and expensive efforts to improve the environment. In this sense, it is not useful to attempt to quantify the developing countries' 'willingness to pay' for improved environmental quality, when their access to merely the basic livelihood essentials typically requires the sacrifice of environmental quality for short-term economic gain. Their ability to pay, or effective demand for environmental quality is so limited under these circumstances that attempts to construct a level of 'willingness to pay' must be speculative at best.

These uncomfortable facts have important implications for the ultimate utility of efforts to quantify assessments of environmental value in the Third World. No matter how complex and sophisticated the price imputation techniques, for instance, the revaluation of tropical forest to include its 'full' environmental value would do little directly to prevent forest destruction, although it might serve to highlight the scale of the problem. Colombia's foreign debt, which requires the country to obtain foreign currency, enables the transnational companies buying valuable hardwoods in protected areas to pose as national saviours, rather than national vandals.

Equity considerations, in this context, are not necessarily a minor element in total utility, as Pearce suggests (Pearce *et al.* 1989: 48), but are often the driving force behind indiscriminate resource degradation, and must be recognized as such. The process of environmental degradation, including the wanton destruction of primary tropical forest, needs to be viewed within the context of highly unequal landholding, which forces poor men and women to colonize the tropical forests and other untitled land. In situations like those of tropical Colombia and Brazil we need to specify greater equity, or the reduction of poverty, as the *primary objective* of sustainable development, before the question of environmental quality can be fully addressed.

It is also essential that we widen the discussion of sustainable development to include the immediate influences of national and regional policies on environmental management decisions taken at the local level. It is at this level that we are least able to provide a clear framework of policy interventions, although a start has been made (IUCN 1988). There is considerable evidence, much of it drawn from the experience of people living within fragile environments, about alternative, more sustainable uses to which resources can be put. In addition, largely because of the work of Pearce and other economists who take the environment seriously, we now have a much better basis from which to conduct environmental accounting within such environments.

These important advances, however, do not imply that the reformulation of environmental policy in developing countries should be confined to an assessment of environmental and economic 'trade-offs', for to do so would mean ignoring other essential points of reference. These include the regional and national political economy of resource use, as well as dimensions of social justice which provide the backcloth against which much environmental degradation occurs. On its own, resource accounting also tacitly endorses a highly ethnocentric, and 'North-biased' view of the development process. Without attention to the analysis of resource use decisions, and the way these are influenced by structures of power and social relations at the community level within the South, we are unlikely to be able to influence the behaviour of people who cut down primary forests in order to make a living.

An approach that is ultimately more successful than these primarily economic views of sustainable development is that taken by the Brundtland Commission's report, *Our Common Future* (Brundtland Commission 1987). Although the economic concept of discounting plays a key role in the report, Brundtland immediately enlarges the compass of the debate about sustainability to include consideration of non-economic factors. *Our Common Future* places the emphasis of the discussion of sustainable development on human needs, rather than on the trade-offs between economic and biological systems. While the future effects of present economic development are a central concern of the report, costs and benefits (both present and future) are assessed not only on economic grounds, but also in political, social and cultural terms.

In fact, Brundtland mapped out a very political agenda for shifting the emphasis of development, for the North as well as the South, without departing from the language of consensus. According to the Commission, 'sustainable development is a process in which the exploitation of resources, the direction of investments, the orientation of technological development and institutional change are all in harmony, and enhance both current and future potential to meet human needs and aspirations' (Brundtland Commission 1987: 46).

One of the important things to notice about the approach taken by the Brundtland Commission is that it regards sustainable development as a policy objective, a methodological approach *and a normative goal*, quite properly the end-point of development aspirations. Many economists acknowledge that normative considerations are important, but few would be prepared to state as unequivocally as does Brundtland that, without normative goals of this kind, improved methodologies and better designed policies will prove unworkable. Brundtland places the responsibility for environmental problems, and for mobilizing the political will to overcome them, firmly in the hands of human institutions and interests. Although the report remains committed to convergence and consensus, rather than divergence and conflict, as a means of achieving sustainable development, the clear implication of Brundtland (and one that has broad appeal in the South, if not the North) is that *unless the political and economic relations that bind the developing countries to the developed are redefined, sustainable development will prove a chimera.*

It is worth noting that some authors, including people like Robert Chambers, who contributed to the Brundtland process, take an even more 'human-focused' approach than that reflected in the report. Chambers argues for using 'sustainable livelihood security' as an integrating concept (Chambers 1988). For Chambers, the sustainability of the resource base makes little sense if it is separated from the human agents who manage the environment. Gordon Conway similarly emphasizes human actors in development. In a series of very influential papers, he argued that 'sustainability [is] the ability to maintain productivity, whether of a field, farm or nation, in the face of stress or shock' (Conway and Barbier 1988: 653). Originally, Conway had been thinking primarily in ecological terms, about the ability of natural systems to cope with system disturbance, and this led him to seek to define a concept which retained the idea of system disturbance, but incorporated a concern for the context of decision-making within which poor rural households operate.

It has been left to the sociologists and anthropologists to take further the discussion of the human agency in sustainable development. In this context, both the participation of people in environmental management at the local level, and the relationship between the implementation of empowering strategies and successful sustainable development, are essential issues to explore.

The multiple dimensions of sustainable development

To establish an adequate conceptual framework within which to explore the issue of participation in sustainable development, we need to identify the multiple dimensions of the concept. There are three dimensions which require our attention: the economic, political and the epistemological.

The economic dimension

As we saw in the discussion of environmental accounting, much of the economic argument has been conducted at the level of present and future anticipated demand, assessing the costs, in terms of foregone economic growth, of closer attention to environmental factors. It was John Stuart Mill, in his *Principles of Political Economy* (1873), who emphasized the idea that we need to protect nature from unfettered growth if we are to preserve human welfare before diminishing returns begin to set in. Malthus had earlier stressed the limits of the carrying capacity of the environment, although his emphasis was on the adverse effects that population pressure would have on consumption, rather than on the impact of environmental degradation itself.

Mill's concern with the environment, which today we would identify as part of the alternative, sustainable tradition of thought, has not been integrated into the mainstream of economic theory during this century. Following Ricardo's much more optimistic assessment of the potential of technology to overcome the limitations of existing resources, the more recent tradition has been to rely on humankind's promethean spirit and ingenuity to enable society to make scientific and technological advances capable of 'putting back' the day in which population growth would begin to overtake available resources.

This optimism was shaken, although not destroyed, by the publication of *Limits to Growth* in the early 1970s (Meadows *et al.* 1972). This influential book argued that natural resources were indeed in short supply, undermining the assumption that humankind could continue to overcome the obstacles placed in its path by nature. The 1970s was a time in which – particularly following the oil price shocks – economic growth endangered the planet, primarily because the clamour for growth had meant the neglect of the environment on which growth was dependent. Twenty years later, the situation in the developed world is different: today we are beginning to be aware that it is the damage to our environment, caused by a heavy dependence on fossil fuels to drive industrial growth, that potentially imperils our ability to continue to support industrial society. The global externalities today, notably the greenhouse effect and the depletion of the ozone layer, are not the product of scarcity but of reckless and unsustainable production systems.

The political dimension

The political dimension of the concept of sustainability comprises two separate but related elements: the weight to be attached to human agency and social structure, respectively, in determining the political process through which the environment is managed; and the relationship between knowledge and power in popular resistance to dominant world views of the environment and resources. In both cases it is useful to draw on a body of emerging social theory which has evolved and gained currency while environmentalism has risen to prominence.

The problem of human agency in relation to the environment is well recognized in the literature, especially by geographers (O'Riordan 1989). It is also a central concern of sociologists, although rarely linked to environmental concerns per se. The British sociologist Anthony Giddens has devoted considerable attention to what he describes as a theory of 'structuration', which would enable us to recognize the role of human beings within a broad structural context in seeking to advance their individual or group interests (Giddens 1984). Giddens notes that 'human agents ... have as an inherent aspect of what they do, the capacity to understand what they do while they do it' (Giddens 1984: xxii). It is their *knowledgeability* as agents which is important. Although Giddens does not apply his ideas specifically to environmental questions, they have clear utility for any consideration of the political and social dimensions of sustainability.

An examination of the ways in which power is contested helps us to explain human agency in the management of the environment, as well as the material basis of environmental conflicts. In this sense it is useful to distinguish between the way human agents dominate nature – what has been termed 'allocative resources' – and the domination of some human agents by others, or 'authoritative resources' (Giddens 1984: 373). Environmental management and conflicts over the environment are about both processes: the way groups of people dominate each other, as well as the way they seek to dominate nature. Not surprisingly, the development, or continuation, of more sustainable livelihood strategies carries important implications for the way power is understood between groups of people, as well as for the environment itself. The 'green' agenda is not simply about the environment *outside human control*; it is about the implications for social relations of bringing the environment within human control.

The second question of importance in considering the political dimension of sustainability is the relationship between knowledge and power, a dimension often overlooked by observers from developed countries when they turn their attention to poorer societies. As we shall see in a moment, the consideration of epistemology in sustainable development carries important implications for our analysis, since it strikes at the cultural roots of quite different traditions of knowledge. It is also important to emphasize, however, that knowledge and power are linked, as Foucault observed in much of his work (Smart 1985; Sheridan 1980). We can, following Foucault, distinguish three *fields of resistance* to the 'universalizing' effects of modern society, and these fields of resistance are particularly useful in delineating popular responses, by the rural poor in particular, to outside interventions designed to manage the environment in different ways.

The first type of resistance is based on opposition to, or marginalization from, production relations in rural societies. This is resistance against *exploitation*, and includes attempts by peasants, pastoralists and others to resist new forms of economic domination, which they are unable to control or negotiate with.

33

The second form of resistance is based on ethnic and gender categories, and seeks to remove the individual from domination by more powerful groups whose ethnic and gender identity has conferred on them a superior political position. In many cases the only strategy open to groups of people whose environmental practices are threatened by outsiders, and whose own knowledge, power and identity are closely linked with these practices, is to seek to distance themselves from 'outsiders' by, for example, reinforcing ethnic boundaries between themselves and others.

Finally, poor rural people frequently resist *subjection* to a world view which they cannot endorse, in much the same way as people in developed countries often confront 'totalizing' theories, such as psychoanalysis or Marxism. In the South, development professionals frequently have recourse to a body of techniques for intervening in the natural environment which are largely derived from developed country experience. 'Environmental managerialism' is one way of describing these techniques. The refusal to be subordinated to a world view dominated by essentially alien values and assumptions marks resistance against subjection. This does not imply that such resistance should necessarily be equated with political struggle, whatever the basis of the resistance itself. Frequently, people who are relatively powerless, because their knowledge-systems are devalued, or because they do not wield economic power, resist in ways which look like passivity: they keep their own counsel, they appear 'respectful' towards powerful outsiders, but they simply fail to cooperate.

The epistemological dimension

Sustainable development is usually discussed without reference to epistemological issues. It is assumed that the system of acquiring knowledge in the North, through the application of scientific principles, is a universal epistemology. Anything less than 'scientific knowledge' hardly deserves our attention. Such a view, rooted as it is in ignorance of the way we ourselves think, as well as of other cultures' epistemology, is less than fruitful. Goonatilake (1984) reminds us that large-order cognitive maps are not confined to Western science, and that in Asia, for example, systems of religious belief have often had fewer problems in confronting 'scientific' reasoning than has the Judaeo-Christian tradition. The ubiquitousness of Western science, however, has led to traditional knowledge becoming 'fragmented' in the South, increasingly divorced from that of the dominant scientific paradigm.

The philosopher Feyerabend, in his influential book *Farewell to Reason*, has distinguished between two different traditions of thought, which can usefully be compared with 'scientific' and 'traditional' knowledge. The first tradition, which corresponds closely to scientific epistemology, is the *abstract tradition*. This enables us 'to formulate statements [which are] subjected to certain rules [of logic, testing and argument] and events affect the statements only in accordance with the rules. ... It is possible to make scientific statements without having met a single one of the objects described' (Feyerabend 1987: 294). He gives as examples of this kind of tradition elementary particle physics, behavioural psychology and molecular biology. In contrast, the kinds of knowledge possessed by small-scale societies Feyerabend would label as *historical traditions*. In these epistemological traditions 'the objects already have a language of their own', and the object of enquiry is to understand this language. In the course of time much of the knowledge possessed by people outside mainstream science, especially in developing countries, becomes encoded in rituals, in religious observations and in the cultural practices of everyday life. In societies which make an easy separation between 'culture' and 'science' such practices can easily be ignored, although they are frequently the key to the way environmental knowledge is used in small-scale rural societies.

It is evident from some of the cases discussed briefly in the later sections of this chapter that any view of epistemology which rests solely on Northern experience will often fail to galvanize opinion among people such as the Brazilian rubber tappers or the Indian women involved in the Chipko movement. What is required is the admission that we are dealing, when we observe local resource management strategies, with *multiple epistemologies* possessed by different groups of people. Furthermore, the existence of global environmental issues, and the reporting of these issues by the media, forces us to consider the links between local epistemologies (all of which have evolved from their own encounter with other systems of thought, and are not fixed, 'traditional' systems) and global systems of knowledge.

THE RURAL POOR AND SUSTAINABLE DEVELOPMENT: OUTSIDE INTERVENTION, INSIDE KNOWLEDGE

The first part of this chapter has sought to extend the definition of 'sustainable development' by enlarging the compass of debate, and considering the dimensions of sustainability which usually lie outside the parameters of most Northern environmental policy interventions. As such it represents a contribution to the still small body of work which has begun to examine the links between local environmental knowledge, political processes and the management of resources (McNeely and Pitt 1985; IUCN 1989; Norgaard 1985). By enlarging the discussion it is hoped that we can begin to get at the texture of 'actually existing' sustainable practices, and thus to make more qualified decisions about the direction that future policy should take. The remainder of the chapter employs the framework of sustainable development outlined above in order to consider the role of external agencies and local knowledge in a more genuinely participatory view of resource management.

Because environmental management in the North utilizes a scientific epistemology, development 'experts' frequently devalue the contribution of local knowledge to environmental planning and policy and, simultaneously, assume that local people should 'participate' in sustainable development. However, it is not clear why or how poor people can retain their knowledge systems, and put them to practical use within development activities, while 'participating' in other people's projects.

Rural people are unlikely to perceive the problems which face them in everyday life as 'environmental problems'. Nevertheless the 'answers' arrived at by the state, and other outside institutions, make assumptions about what is beneficial for people, and ways in which the environment can be more effectively managed (Blauert 1990). In fact, the approaches of outside agencies frequently address the problems of the agencies themselves, rather than those of the rural poor or their environments. To most poor people in rural areas, for whom daily contact with the environment is taken for granted, it is difficult, if not impossible, to separate the management of production from the management of the environment, and both form part of the livelihood strategy of the household or group. It is increasingly recognized by many development agencies, notably NGOs working in developing countries, that the sectoral, 'single problem' approach to policy and planning undertaken by most official bodies prevents a workable assessment of sustainable development options.

The current call for more participatory approaches to local-level environmental management stems from the failure to recognize the importance of popular participation in influential reports such as those of Brandt (1980) and Brundtland (1987) as well as the original World Conservation Strategy document (IUCN 1980). It also reflects the acknowledgement that national governments are less likely to ignore international opinion when it is buttressed by popular, grassroots support.

The call for more participation also reflects a third important variable: during the 1970s and 1980s an influential body of knowledge, along with new methodological interventions, stressed the importance of capturing the knowledge of poor people themselves – through farming systems research, agroecology and 'rapid rural appraisal' techniques. However, the cultural and political aspects of these gains in understanding received almost no attention. Social structure and political action remained essentially outside the map of development policy at the micro-level, and were given scarcely any attention in discussions of the natural environment.

The problem of rural poverty and the environment has frequently been posed in terms of available and appropriate technologies, while more reflexive, more iterative ways of working with rural people in developing countries were confined to the relatively 'marginal' concerns such as community development. Anthropologists, for example, frequently found unlikely allies in ecologists, whose negative experience of large-scale development projects echoed their own (Ewell and Poleman 1980).

It often appeared as if the larger the financial commitment of an organization to 'development' goals, the smaller was the commitment to discovering how to assist the empowerment of the poor, drawing on their knowledge, their priorities and their politics. One of the consequences, with which we grapple today, was that most environmental knowledge, like environmental management, is handed down from the First World to the Third, from large development agencies to the supposed beneficiaries of change.

The report of the World Commission for Environment and Development, *Our Common Future* (Brundtland Commission 1987) served to set the agenda for recent thinking about the environment and development. Despite its trenchant analysis, accessible style and clear exposition of the issues, the Brundtland Commission has relatively little to say about popular participation in environmental management at the local level. Other than a few short, but useful, sections on participation the Commission's report says little about local empowerment until the conclusion, in which, after a long account of the international measures required to achieve more sustainable development, a short section on popular participation is included:

> progress will also be facilitated by recognition of, for example, the right of individuals to know and have access to current information on the state of the environment and natural resources, the right to be consulted and to participate in decision making on activities likely to have a significant effect on the environment, and the right to legal remedies and redress for those whose health or environment has been or may be seriously affected:
>
> (Brundtland Commission 1987: 330)

Despite the fact that these points are not elaborated in the Report, and popular involvement in environmental management gets only the most cursory treatment, these few phrases represent a commitment of immense value, which deserve to be taken seriously by the international community and national governments. Suddenly the issue of sustainable development is linked to human rights, and these rights are specified in terms of 'their' right to know and be consulted. Participation, it is implied, is not simply a means of ensuring the efficacy of 'our' development (via more attention to factors such as the creation of employment) but a means of ensuring their sustainability through the possession of the rights without which it cannot be achieved.

Evidence for greater attention to participation, and with it poor people's rights in the environment, can be gleaned from the first draft of the *World Conservation Strategy for the 1990s*, prepared by IUCN, UNEP and the WWF (IUCN 1989). This document goes some way to redressing the lack of attention to people in the original *World Conservation Strategy* (IUCN 1980). The discussion of 'policy, planning, legislation and institutions' (pp. 137–44) pays particular attention to the obligations which a more sustainable development strategy places on governments, to consult them, to facilitate their participation in decisions, and to make information available to them. It also recognizes that 'special attention should be given to participation by women and indigenous peoples', which should be provided for by governments and intergovernmental agencies (p. 138).

The final section of the document gives considerable attention to local strategies for sustainable development, arguing that local communities should be given the opportunity to prepare their own sustainable development strategies 'expressing their views on the issues, defining their needs and aspirations, and formulating a plan for the development of their area to meet their social and economic needs sustainably' (p. 156). This should be undertaken, like the regional and national strategies to which it would contribute, on the basis of consensus. Achieving 'a community consensus on a future for an area' would require consultation and agreement with other, non-community interests, as well as 'a forum and process through which the community (itself) can achieve consensus on the sustainable development of the area' (p. 157).

In practice, however, in most developing countries local-level environmental management will be left to understaffed, under-funded and underesteemed enforcement agencies. The new World Conservation Strategy recognizes that legislative changes will be necessary before sustainable development strategies can be implemented with any success, but it attempts no analysis of the forces at the local, national and international level which would need to be pressed into service to ensure that legislation is enforced and local management decisions are implemented. This document, in fact, shares the assumptions of much discussion of 'participation', which is predicated on the presence of a social consensus that, in practice, rarely exists, especially in the most threatened parts of developing countries. Unless we analyse specific power structures in relation to the environment, we are in danger of being far too sanguine about the potential of negotiation and agreement. We are in danger, in fact, of drowning in our own rhetoric rather than identifying the underlying political processes whose understanding would facilitate the formulation of better environmental policy.

CONFLICTS OVER RESOURCE MANAGEMENT: FORMS OF RESISTANCE

Table 2.1 sets out some of the important variables for an analysis of conflicts over resource management at the local level. It must be emphasized that in the cases described the resources in question are

Table 2.1 Conflicts over resource management: forms of resistance

	Choices for resource utilization	Political demands	Points of tension and resistance	State/external intervention
Chipko (Shiva and Bandyopadhyay 1986; Guha 1989)	(1) Forest conservation (2) Commercial logging	Respect for traditional forest uses	Peaceful non-cooperation (*satyagraha*)	Indian Government intervention
Brazilian rubber tappers (LAB 1990; Hecht and Cockburn 1989)	(1) Sustainable forest extraction (2) Ranching	Conservation reserve	Forest clearing, federal government support	Brazil-wide solidarity groups International ecological awareness
Tropical colonists (Bolivia) (Redclift 1987)	(1) Sustainable farming system (2) Commercial rice (cultivation/land engrossment)	Land titles Institutional support	Disputed land ownership Migration Economic policy	Land reform Cocaine surveillance
'Freelance' logging (Choco, Colombia)	(1) Contracted 'logging' for TNC (2) Community stewardship		Individual livelihood strategy v INDERENA	INDERENA military base

heavily contested, and the conflicts surrounding them have drawn in both national and international interest groups. Many conflicts over local resource management in developing countries lack the heavily politicized nature of the Chipko or Brazilian rubber tappers' disputes, which have attracted media attention and become the focus for alternative development agendas. Nevertheless, these conflicts, and others such as the cases of Bolivian frontier colonists, and freelance logging in the Choco of Colombia do illustrate the inadequacy of environmental interventions which proceed on the assumption of existing consensus, and in ignorance of the social and political struggles which lie behind environmental disputes.

The conflicts between Chipko activists in India, logging companies and the Indian Government are well known and have been exhaustively discussed in the literature (Bandyopadhyay, this volume; Guha 1989; Shiva and Bandyopadhyay 1986; Kunwar 1982). Similarly, the struggle of the Brazilian rubber tappers in the Amazon to establish their rights to use the forest in a sustainable way has received extensive coverage, notably since the murder of the rubber tappers' leader, Chico Mendes. The struggles of the rubber tappers have reached the world stage, especially through the press and television, but the precise circumstances of the conflict require some explanation (Schwartzman 1989; Hecht and Cockburn 1989; Hecht 1989).

According to Schwartzman (1989) there are approximately 1.5 million people in the Brazilian Amazon who depend on the forest for their living. Of these, about 300,000 are engaged in the sustainable harvesting of wild rubber. In fact, most rubber tappers, like other sectors of the forest population, are involved an several activities other than their main cash-earning occupation: they cultivate small gardens planted with rice, beans and manioc, keep animals and hunt in the forest. They also cultivate and manage fruit trees, palms and other forest species. The rubber tappers' production system 'appears to be indefinitely sustainable. Many rainforest areas have been occupied by rubber tappers for over sixty years, and some families have been on the same holdings for forty or fifty years, yet about 98% of each holding is in natural forest' (Schwartzman 1989: 156).

The diversity of sources of income is reflected in various aspects of the rubber tappers' culture: their diet is much more varied than that of most urban groups; their average cash income, although not large, is equivalent to twice the Brazilian minimum wage; and their awareness of the links between their livelihood and the maintenance of ecological diversity has enabled them to present their case as a convincing one of sustainable development. Any suspicion that their case has received special attention needs to be set against the fact that most other economic activities in the Amazon receive much higher subsidies, and are usually accompanied by disastrous effects.

In terms of local resource management, the interest in the rubber tappers' activities lies in two important issues. First, unlike much of the conservationist response currently being urged on governments in the South, the extractive reserves advocated by the rubber tappers are not simply another culturally alien 'management strategy' urged on unwilling, or oblivious, local people. The idea of extractive reserves is an organized initiative directly undertaken by Amazonian grassroots groups and sympathetic national organizations, designed to change the course of official regional development policy for the benefit of local people. Because the extractive reserve concept was created by a social movement, it does not depend for its effective implementation on government agencies far removed from Amazonian reality. Forest communities have put their own model before the government and multinational lending institutions as a potential strategy for consideration within a wider context of sustainable development.

Second, although locally sustainable, the rubber tappers' activities also produce a surplus which finds its way to the larger society: this is a movement that is not only locally initiated, but is also one that generates momentum outside the immediate domain of the *seringueiros* (rubber tappers).

The other two cases presented in the table are less well known. The tropical colonists referred to in the third case are largely migrants from the Bolivian Andes who migrated to the lowland province of Santa Cruz in the 1960s and 1970s, in search of land. These migrants have concentrated on growing rice for the market, but the difficulties associated with cutting down the forest, and the insecurity of the market for rice has also led some of them to explore (with official encouragement from some quarters) a more mixed farming system, comprising rice, perennial crops and small-scale animal production. The problems of managing a more sustainable system in an area where conflicts over land are compounded by contraband traffic and the cocaine trade are outlined in Redclift (1987).

The final case is illustrated by the conflict between INDERENA, a Colombian environmental agency, and the people living in the area of the Choco, a reserve situated on the tropical Pacific coast of Colombia. These people were able to receive $10 a cubic metre for hardwoods cut from the forest reserve with chain saws loaned by a transnational company operating in Colombia, Cartón de Colombia. Each load of hardwoods had to be taken by sea, on a home-made raft, out into the Pacific and on to the port of Buenaventura. There was considerable resentment in the area at the attempts, usually futile, of the INDERENA staff to prevent the cutting of wood in this way. For the people involved in illegal cutting, the activity represented an essential livelihood strategy, and there was no shortage of men willing to take the place of those who did not survive the dangerous sea journey. It is also worth mentioning that Cartón de Colombia is a major sponsor of environmental activities in Colombia (including the conference organized by INDERENA that I was attending).

The tragedy of hardwood logging in the Choco, even on the relatively small scale practised by 'freelance' colonists, is that with sufficient official support, sustainable alternatives for the area could be implemented. It is thought that the Choco possesses 'perhaps the most diverse plant communities in the world and extremely high levels of local, as well as regional endemic species' (Budowski 1989: 274). Two sustainable strategies, in particular, have attracted attention, because they would make no serious inroads into the region's ecological diversity but would enable large numbers of people to make a decent livelihood. First, food production could be concentrated on the rich alluvial river banks where, together with agroforestry combinations, larger populations could be supported. Second, if sustainable forestry schemes were promoted, especially in the swamp and secondary forests, numerous opportunities would open up for settlers in the region. The potential for the sustainable yield of freshwater fisheries in the area is even greater (Budowski 1989: 276). Finally, it is clear that the ecological value of the Choco is so great in global terms, that international efforts to promote local research activities, and to promote research stations within the region, linked to local communities, would bring about huge advances in our knowledge, especially of better-drained forested areas.

Each of the cases referred to in the table is related, along the horizontal axis, with four dimensions of the conflict: the alternative choices available for resource utilization in the area; the political demands of the participants in the various social movements; the points of tension and forms of resistance employed during the conflict; and the form of outside, state intervention to mediate the situation. In the cases of the Chipko movement and the rubber tappers, the conflict surrounds the defence of an existing, sustainable resource use or livelihood. In the case of the Bolivian colonists, a sustainable alternative to existing resource uses was available, but the incentives to make it attractive to people did not exist. The framework of policy measures and incentives in the Santa Cruz region of Bolivia favoured short-term calculations of profit over longer-term considerations of sustainability, although the risks carried by involvement with the market also threatened profitability for the colonist farmers. In the case of the Choco, the individual's logging activities were undertaken independently of any community structure: individual livelihood opportunities were pursued in opposition to the formal, legal framework, but 'supported' by a powerful transnational corporation.

The points of tension for each of the conflicts are different, and the interest of outsiders in the conflict vary widely, especially in terms of the commitment of the state to intercede on behalf of one group rather than another. In addition, it is impossible to view these conflicts as divorced from wider patterns of influence on the governments concerned, and in a more general sense in reshaping our awareness of the urgency of ecological issues. Although the local agents seem remote from most people, not only in the North, but also from the population of Indian or Latin American cities, their struggles provide evidence of the interdependence of both economic forces and power relations.

Before considering the need to examine these power relations in more detail, it is worth reflecting on the potential value of an approach to resource management which explicitly recognizes the importance of popular participation. It is clear from these and other similar cases that forms of political activity over the environment vary widely: we should not expect popular participation to follow a single trajectory. Second, it needs to be emphasized that in the course of conflicts over natural resources, new priorities and development opportunities are opened up and brought within the compass of popular discourse. The determination of development trajectories is not confined to the offices of experts working for the World Bank or of academic observers; they are worked out in the heads of the subjects themselves. Third, resistance to the 'totalizing' effects of incorporation, even at the geographical periphery, into modern society can lead to the formulation of demands which have to be negotiated with governments and international interests.

A commitment to a more democratic discourse on the part of governments or the international development community, however, is only one of several possibilities whose probability depends, critically, on the role of supportive groups and interests, including NGOs, international pressure groups, and classes within the society itself. The mediation of conflicting demands and their peaceful resolution might be the outcome of resource conflicts, but it is unhelpful to assume that general agreement of this kind can be found, and that better environmental management is virtually impossible without it. The discussion of environment and development by international agencies frequently fails to identify the alternatives to consensus, or the role that the recognition of conflicting interests can play in policy formation. The more closely we examine conflicts over resource management in developing countries, the more we need to pay attention to the political and social mechanisms through which interests in the environment are channelled and expressed. It is therefore to this question, for so long ignored in discussions of resource management, that we turn in the final section.

CONTESTED RESOURCES: POWER, RESISTANCE AND SOCIAL CHANGE

At the beginning of this chapter it was suggested that conflicts over the environment could be analysed in terms of three dimensions: the economic, the political and the epistemological. It was argued that power and resistance were complementary aspects of the same strategic situation. Further, it was suggested that the way the environment was viewed in different cultures corresponded with distinct epistemological traditions of thought. We should not assume that knowledge, whether 'local' or 'scientific', could be easily separated from ways of behaving, ways of managing resources, or ways of expressing resistance towards the attempts of others to manage resources.

The current rethinking of mainstream economics, and the greater incorporation of environmental considerations which is highly influential within some development agencies, is helping to fashion

45

a tool for policymakers in the North, but there are limitations to the heuristic possibilities which such techniques provide. Any serious discussion of participation in resource management – and any analysis of the problem – needs to consider the full range of demands which the management of natural resources involves. We should not pursue better resource management within an apolitical, normative conceptual framework of our own making. We need to take seriously the resource politics of people in the South, especially since their own political consciousness is forged through contact with external development agencies, planning institutions and policymakers.

The articulation of demands governing the use of natural resources inevitably means the exercise of power, and resistance to it. It should come as no surprise, then, to find that environmental demands affect the content of social relationships, as well as the form. They bring new social relationships into being, and with them new power relations, many of them uncomfortably like those they have superseded. In some cases a radical break is achieved, through which existing relations are democratized or opened up, but there is no guarantee that the new relations of power that are established will be more stable. Every strategy of confrontation dreams of becoming a relationship of power, of finding a stable mechanism to replace the free play of antagonistic forces. However, there is no guarantee in history that this will happen. As we have seen, frontier colonists in Brazil and villagers in India do not demand the end of the State or law, but insist instead on respect from the government for rights which are enshrined in tradition, as well as law.

The approach I have outlined to power relations can be used in exploring the contests between human agents over environmental resources. For example, peasant movements may be contained by a chain of state agencies through which power relations are deployed and reformulated (Harvey 1989). By identifying the weaker and stronger points in this chain, movements can apply pressure to break the former with the goal of eventually breaking the latter. If we begin by identifying the most important points of tension in local society, and the conflicts they generate, we can observe how the specific application of power is resisted and transformed, how new tactics are introduced and how traditional mechanisms are abandoned.

Bearing these points in mind we can propose a set of questions which can help us establish better methodological guidelines for the comparative analysis of micro-political change in relation to the environment. We can usefully compare the different ways in which groups seek to control and manage resources, and the concrete implications of these strategies for external agencies whose remit is to help channel and facilitate the expression of local demands. We need to look closely at the way in which different groups establish power relations through their control over resources, and the way in which these power relations change over time. In this respect, the following sets of questions can be posed:

1 How do legal and institutional changes limit or enable groups to engage in *particular forms of political action over the environment*? Which groups have most successfully integrated their own micro-strategies with wider strategies shared by other members of the society? As it becomes clear that different groups in the wider society acquire different notions of 'sustainability', carrying implications for their own political action, it becomes more urgent that local demands are linked to wider social resistance.

2 How does the recomposition of power relations affect the political priority given to more sustainable resource management? Do new strategies of political mediation, or domination, make certain policy alternatives less feasible, while opening up new ones? How do local agents view the constraints and opportunities which changing resource uses make possible? Are they able to carry their alternative vision of sustainability, their 'concrete utopia', into the organs of the state itself?

3 How do struggles over resources shape the paths of different social groups? Do they channel environmental demands into the institutional arena alone, or do they engage groups in confrontations which highlight basic divisions within the wider society? What are the effects upon NGOs and governmental agencies of intervention to secure long-term environmental demands? Is it the case, as the Brundtland Commission hoped, that more contact between development agencies serves to bring forward the urgency of environmental priorities within policymaking circles?

These considerations are offered as a contribution to the resolution of some of the conceptual and methodological issues which surround local resource management. By identifying the points of tension in local systems of power, and comparing their implications for different groups, often possessed of different epistemological systems, we will be able to highlight the changes through which the environment becomes the object of economic, social and political dispute. The lessons of the past and of the present are central to any strategy of resistance and liberation, but it is up to us to undertake the necessary analysis, and to place it in the hands of those disempowered by the development process.

REFERENCES

Barbier, E. (1989) *Economics, Natural-resource Scarcity and Development*, London: Earthscan.

Bartelmus, P. (1987) *Environment and Development*, London: George, Allen & Unwin.

Blauert, J. (1990) 'Autochthonous approaches to rural environmental problems: the Mixteca Alta, Oaxaca, Mexico', University of London Ph.D. (Wye College).

Brandt Commission (1980) *North–South: a programme for survival*, London: Pan Books.

Brundtland Commission (1987) (see WCED 1987).

Budowski, G. (1989) 'Developing the Choco region of Colombia', in John O. Browder (ed.) *Fragile Lands of Latin America: strategies for sustainable development*, Boulder: Westview Press.

Chambers, R. (1988) 'Sustainable rural livelihoods: a strategy for people, environment and development', Institute of Development Studies: University of Sussex.

Conway, G. and Barbier, E. (1988) 'After the green revolution: sustainable and equitable agricultural development', in D. Pearce and M. Redclift (eds) *Futures*, 20 (6): 651–78.

Devall, B. and Sessions, G. (1985) *Deep Ecology: living as if nature mattered*, Layton, Utah: Peregrine Smith.

Ewell, P. and Poleman, T. (1980) *Uxpanapa: agricultural development in the Mexican tropics*, Oxford: Pergamon.

Feyerabend, P. (1987) *Farewell to Reason*, London: Verso.

Giddens, A. (1984) *The Constitution of Society*, Oxford: Polity Press.

Goonatilake, S. (1984) *Aborted Discovery, Science and Creativity in the Third World*, London: Zed Books.

Guha, R. (1989) *The Unquiet Woods: ecological change and peasant resistance in the Himalaya*, New Delhi: Oxford University Press.

Harvey, N. (1989) 'Corporatist strategies and popular responses in rural Mexico: state and opposition in Chiapas, 1970–1988', University of Essex Ph.D.

Hayter, T. (1989) *Exploited Earth: Britain's aid and the environment*, London: Earthscan/Friends of the Earth.

Hecht, S. (1989) 'Chico Mendes: chronicle of a death foretold', *New Left Review*, 173: 47–55.

Hecht, S. and Cockburn, A. (1989) *The Fate of the Forest Developers, Destroyers and Defenders of the Amazon*, London: Verso.

IUCN (International Union for the Conservation of Nature and Natural Resources) (1980) *World Conservation Strategy: Living Resource Conservation for Sustainable Development* (with UNEP and WWF), Gland, Switzerland: IUCN/UNEP/WWF.

IUCN (International Union for the Conservation of Nature and Natural Resources) (1988) *Economics and Biological Diversity: guidelines for using incentives*, Gland, Switzerland: IUCN.

IUCN (International Union for the Conservation of Nature and Natural Resources) (1989) *World Conservation Strategy for the 1990s* (with UNEP and WWF), Gland, Switzerland: IUCN/UNEP/WWF, draft.

Kunwar, S.S. (ed.) (1982) *Hugging the Himalaya: the Chipko experience*, Gopeshwar.

LAB (Latin American Bureau) (1990) *Fight for the Forest: Chico Mendes in his own words*, London: Latin American Bureau.

McNeely, J. and Pitt, D. (eds) (1985) *Culture and Conservation: the human dimension in environmental planning*, London: Croom Helm.

Meadows, D.H., Meadows, D.L., Randers, J. and Behrens, W. (1972) *The Limits to Growth*, London: Pan.

Mill, J.S. (1873) *Principles of Political Economy*, London: Parker, Son and Bourn.

Norgaard, R. (1985) 'Environmental economics: an evolutionary critique and a plea for pluralism', *Journal of Environmental Economics and Management*, 12 (4): 382–94.

O'Riordan, T. (1989) 'The challenge for environmentalism', in R. Peet and N. Thrift (eds) *New Models in Geography* (volume 1), London: Unwin Hyman.

Pearce, D., Markandya, A. and Barbier, E. (1989) *Blueprint for a Green Economy*, London: Earthscan Publications.

Redclift, M.R. (1987) *Sustainable Development: exploring the contradictions*, London: Methuen.

Schwartzman, S. (1989) 'Extractive reserves: the rubber tappers' strategy for sustainable use of the Amazon rainforest', in John O. Browder (ed.) *Fragile Lands of Latin America: strategies for sustainable development*, Boulder: Westview Press.

Sheridan, A. (1980) *Michel Foucault: the will to truth*, London: Tavistock.

Shiva, V. and Bandyopadhyay, J. (1986) 'The evolution, structure and impact of the Chipko movement', *Mountain Research and Development*, 6 (2): 133–42.

Smart, B. (1985) *Michael Foucault*, London: Tavistock/Ellis Horwood.

WCED (World Commission on Environment and Development) (1987), *Our Common Future*, Oxford: Oxford University Press.

World Bank (1987) *Environment, Growth and Development*, Development Committee Paper 14, Washington, DC.

World Bank, (1988a) *Environment and Development: implementing the World Bank's new policies*, Development Committee Paper 17, Washington, DC.

World Bank (1988b) *The World Bank and the Environment*, internal discussion paper, Washington, DC.

3

FOUNDATIONS FOR SUSTAINABLE DEVELOPMENT: PARTICIPATION, EMPOWERMENT AND LOCAL RESOURCE MANAGEMENT[1]

Jessica M. Vivian

INTRODUCTION

Much of the action taken by development practitioners to address local-level environmental problems in the Third World consists of projects such as tree-planting schemes, soil-bunding efforts or improved irrigation management strategies which seek to establish resource use at sustainable levels for selected target areas. In spite of occasional suggestions that broader-level national or international policies should be formulated with the aim of making natural resource management concerns an integral part of economic and social policy (e.g. Warford 1987), this type of approach remains dominant. It is therefore not surprising that the current discussion of environment and development issues often mentions 'people's participation' as a prerequisite for successful 'sustainable development'. Resource management projects, as currently implemented, depend heavily on broad-based cooperation and collaboration because they often rely on the combined actions of individuals which – whether such actions be planting trees or refraining from overfishing – by their nature cannot easily be coerced or enforced. The willingness of people to undertake the required activities – what is conventionally understood as their 'participation' – is therefore essential for the success of these projects. Participation in resource management in the Third World, however, should be understood as a much broader concept – it takes many forms, and is not limited to people's contribution of time and labour to externally developed initiatives.[2] This chapter discusses ways in which a more thorough understanding of the range of activities which constitute people's participation in local-level environmental activities – from the development of indigenous resource management systems to resistance to destructive external initiatives – can be used to form the basis of a more constructive approach to sustainable development.

The analysis contained in this chapter follows from some of the work undertaken within the UNRISD research programme on sustainable development and participation in resource management, which explores, among other things, the dynamics of local-level initiatives concerned with environmental degradation and traditionally sustainable resource management practices. Although definitive findings from this programme are not yet available, the research undertaken to date has indicated a number of areas in which the standard interpretation of the dynamics of the process of localized environmental degradation can usefully be re-examined. This chapter explores the issues raised by the research and the insights gained in the process. It opens with a discussion of the prevailing approaches to environmental problems undertaken by the development community, suggesting two areas in which a broader understanding of 'participation' can contribute to the formulation of more productive solutions. It then briefly defines 'sustainable development' as it is used here, and discusses the utility of this concept. The chapter next discusses issues connected with the continued viability of traditional resource management systems, including population pressure, the effect of changing economic structures, common property and human rights issues.

The following section discusses popular initiatives which have affected local-level environmental issues, both in the form of organized participatory activities and social protest movements, and the potential which such initiatives have for arresting or reversing environmental degradation. It is argued that these activities, even those which have evolved precisely to oppose outside developmental interventions, have very important implications for the formulation of more effective sustainable development strategies.

Finally, the chapter examines the question of the apparent linkages between poverty and environmental degradation in the Third World in the light of the issues raised by the research. It is argued that although in certain cases poverty clearly aggravates

51

processes of degradation, an analysis positing a simple linkage between these two is incomplete, and unhelpful in policy terms, without the inclusion of the concept of empowerment.

PARTICIPATION AND CONSERVATION PROJECTS: SOME PROMISING APPROACHES

To the extent that current and future environmental problems can be arrested or reversed through the types of rehabilitation projects usually sponsored by outside donors, or by protection measures taken by governments, research on ways to increase local cooperation with environmental projects is useful. However, evidence is mounting that the targeted project approach to environmental problems, though often well-intentioned and very valuable within a limited scope, will not be sufficient to solve the environmental problems facing the South today. The problems are too widespread, and too deeply entrenched, to be entirely solved with the disparate, sometimes haphazard and usually very localized palliative remedies presently offered. For example, despite the organization, in response to Ethiopia's agricultural crisis, of one of the largest soil rehabilitation projects in the world, Ethiopia's highlands continue to erode: the scope and financing of the project, although massive compared with similar efforts elsewhere, are still far below what would be needed to make a real impact on the environmental problems of the country. As a result, the effects of the conservation project are only evident in small parts of the highlands, while in the isolated areas away from the main roads, where the majority of the farmers live, there are few conservation activities at all (Ståhl this volume). Government environmental protection programmes have similarly limited impacts, due to the often-discussed problems of underfunding, lack of political support, and lack of institutional and technical capacity. The Ethiopia case is only one example which illustrates the limitations of discrete projects for solving the problems of development. As Sithembiso Nyoni writes, 'no nation in the world was developed by projects alone, let alone projects based on borrowed models' (Nyoni 1987: 52).

The fact that environmental degradation in the Third World is commonly perceived as a crisis in the 'sustainable development' literature contributes towards the prevalence of corrective-type projects. A crisis seems to call for immediate and direct measures, and, as Adams (1990) argues in a discussion of development policies, often favours 'firefighting' approaches rather than discussions of deeper ills, and the treatment of symptoms rather than causes. In addition, a project-oriented approach can seem, from the point of view of donors, to be the most practical. Results are visible and measurable, and impact can usually be demonstrated and success stories reported.

Perhaps a third reason for the prevalence of this approach, and a somewhat more troubling one, is that it can reflect, to a greater or lesser degree, a perception that rural dwellers of the Third World need to be 'taught' about the importance of environmental conservation. The *World Conservation Strategy*, for instance, lists 'the lack of awareness of the benefits of conservation and of its relevance to everyday concerns' (IUCN 1980) as one of the problems to be overcome before sustainable development can be attained. The document calls for addressing this problem by public education on environmental issues, and more community involvement in conservation projects. In fact, of course, many rural Third World communities have practised environmental conservation for centuries, and it has much more often been industrialized populations which have had to relearn the value of the environment.

Projects necessarily form the basis of much development work, no matter how 'development' is defined. However, the project approach to sustainable development, as now standardly conceived, must both be improved and supplemented by a greater understanding of the grassroots-level concerns and activities related to the environment if the environmental problems of the South are to be overcome. At least two directions show promise, and have attracted growing attention, although they have not as yet been the subject of much sustained empirical research. These alternative approaches are both based on 'people's participation' – but on 'participation' defined in a much more fundamental sense than that now commonly used in the environmental literature. True popular participation goes much beyond the mere provision of labour and other inputs into projects initiated from outside the community; it involves decisions being taken and plans being formulated on the local level. In the context of development, as Barraclough (1990) points out, increased popular participation is necessarily a confrontational process, as the development goals of the elite normally preclude increased involvement of the poor in resource management decisions. The working definition resulting from the UNRISD research programme on popular participation in development highlights both the process and the conflict inherent in participation, which is referred to as 'the organized efforts to increase control over resources and regulative institutions in given social situations, on the part of groups and movements of those hitherto excluded from such control' (Pearse and Stiefel 1979: 8).

The first of the two approaches to sustainable development which will be discussed in this chapter involves the increased recognition of 'traditional' resource management practices, an analysis of the value of such practices under current and future conditions, and an assessment of ways either to ensure that sustainable practices are maintained, or to adapt the most viable of them for use in different economic, social or environmental contexts. The second approach involves incorporating the concerns, goals and activities of local grassroots organizations and social movements into externally assisted projects, in such a way that such projects become self-sustaining and, more importantly, self-replicating without additional external promotional efforts. A more thorough understanding of the ways in which people participate in resource management is necessary for the successful development of either of these approaches.

SUSTAINABLE DEVELOPMENT

Defining and refining the concept of 'sustainable development' has become a common exercise. Such an exercise, however, remains a necessary preface to an analysis which utilizes this term, because of the plethora of meanings and emphases, and therefore the diverse implications of this commonly stated objective. Sustainable development is most often defined as 'development that meets the needs of the present without compromising the ability of future generations to meet their own needs' (WCED 1987: par. 2.1). This definition leaves a good deal of room for manoeuvre: it does not specify whose model of development should be followed, nor who will determine the economic, social or biological needs of the present or of future generations.

Redclift (this volume) argues for an approach to sustainable development which emphasizes human needs, and which incorporates an understanding of the political and epistemological dimensions of the concept, as well as the economic ones. In many situations in the Third World, improved living levels are dependent to a large extent upon increased consumption of resources. Therefore, this approach to sustainable development necessarily implies that present levels and methods of resource exploitation should not degrade the environment to the point that resource availability in the future will decline, unless this decrease in resource yields can be compensated through resource 'imports'. It is important to recognize, however, that the fulfilment of human needs also depends on environmental factors unrelated to economic or physical resource yields, including the availability of clean air and adequate living space, and, in many circumstances, people's ability to maintain a spiritual, cultural or aesthetic relationship with their environment. Ecosystem conservation, therefore, also plays a part in sustainable development. Thus this approach avoids the two extremes of the current usage of sustainable development: it is specifically not intended that 'sustainable development' should imply that only those natural resources which can be shown to provide a positive yield in a benefit/cost calculation should be protected. Neither is it implied that the resources of the South should, in the name of the good of the planet, fall under the moral jurisdiction of the North.[3]

Redclift also notes the lack of analytical precision of the term 'sustainable development'. Despite the range of meanings attributed to it, however, it is important not to dismiss this concept as a fashionable yet vacuous fad. The very fact of the wide appeal of the concept means that it has had important implications for the direction that development efforts have taken, and for the programme of work on the environment currently gathering momentum in the development community, as well as on national and local levels. The current coincidence of interest in sustainable development emerges from developmentalists' increasing recognition of the importance of preserving natural resources if development is to continue; and conservationists' growing acceptance that, without development, preservation is not possible.[4] In addition, those concerned with local empowerment, indigenous people's rights, or other human rights issues have recognized that, because the environment is often a very local issue, sustainable development has useful connotations for them as well.

However, given that there are deeply entrenched differences in the understanding of sustainable development, the current unity of purpose in working towards this ill-defined goal is likely to dissolve as more concrete (as opposed to conceptual) decisions need to be taken, and trade-offs made. The alliance based on the flexible concept of sustainable development is not inherently stable or, indeed, mutually beneficial to its members, and will inevitably be strained as divergent interests become more clear (Hawkins and Buttel 1990). However, because of the usefulness of the coalition itself in fostering dialogue and cooperation between the different groups, the concept of sustainable development serves a purpose, and should not be rejected either on the basis that it has been co-opted by mainstream economists or ecologists (e.g. Thrupp 1989; Rees 1990), or that it is too amorphous to be useful.

In defining sustainable development in terms of human needs, and focusing on ways in which local-level participation (which includes the involvement of local people in defining the goals to be attained) can form the basis of more successful approaches to reach this goal, a range of important issues is raised. These include the roles of the state and the international development community in determining policy which affects the environment, the mechanisms by which such policies are influenced, the impact of large-scale environmental destruction on the options available to small-scale resource managers, and the role played by systemic and structural factors in influencing the outcome of a number of environmental problems. These issues inform the argument of much of this chapter, although their detailed and systematic analysis are beyond its scope. This approach produces fresh insights into some of the standard interpretations of the sustainable development conventional wisdom, including the viability of traditional resource management systems, the dynamics of common property management, the relationship of population growth to environmental degradation, and the large-scale potential of small-scale popular environmental movements.

TRADITIONAL RESOURCE MANAGEMENT SYSTEMS

People who rely very immediately on natural resources for their livelihood, if they have been successful in establishing a sustainable mode of production, have typically developed methods to ensure the conservation of their environment. Such indigenously developed resource management practices are commonly referred to as 'traditional', although the length of time they have been operable ranges from a few years to millennia, and although they are not static, but are constantly evolving. In general, these methods are more explicit and more formalized in situations where resources are very scarce, such as in arid lands, although implicit rules governing resource use exist as well in situations of relative abundance. On the community level, resource management systems have generally been more evident among the disadvantaged and rural dwellers than among the urban rich, simply because means of livelihood other than direct resource exploitation are less readily available to the former groups.

Such traditional resource management systems are important to examine in more detail in the context of the search for sustainable development. In spite of the inherent limitations of many such systems, and the external and internal pressures to which they are subject (discussed further below), traditional systems have remained not only viable, but also active in many parts of the world. Where still extant today, these systems involve elaborate social, technological and economic mechanisms to safeguard resources.

There are numerous descriptions, for instance, of religious or spiritual significance being attached to certain plants or animals, which are thereby protected. A particularly striking and well-documented example comes from India, where the religious beliefs held by the Bishnoi community have prohibited killing animals or cutting green trees since the fifteenth century. Today, Bishnoi land is said to be a green and flourishing area in the midst of the surrounding Rajasthan desert (Sankhala and Jackson 1985). There are many similar examples of centuries-old environmental reserves, specifically declared as such (e.g. Draz 1985; Farvar 1987). More common, however, are customs prohibiting the exploitation of particularly useful species, such as the peepal tree in Asia or the baobab in Africa, or allowing the harvesting of animals or plants only at certain seasons or otherwise under conditions which minimize damage to their reproductive potential (e.g. Gadgil 1985; Tobayiwa 1985).

Social controls have also been developed in many communities explicitly to regulate resource use, and to ensure that the environment is managed sustainably. The intricate mechanisms governing pastoralists' grazing patterns, and the intimate environmental knowledge upon which such mechanisms are based, have been well documented (e.g. Lane this volume). Herds are moved according to land use rules which prevent either the most productive or the most drought-resistant lands from being overgrazed. Social convention similarly governs the use of water in the communal irrigation management systems which have existed for centuries in several parts of Asia (Farvar 1987; Yabes this volume), while means of restricting use rights over marine, agricultural and forest resources have enabled communities in various parts of the world to sustain their resource base (see e.g. Polunin 1985; Baines 1989; Moorehead 1989; Diegues this volume).

In addition to communities which have well-defined and explicit rules governing resource use, there are many situations in which resource use regulations only become evident to outside observers when overexploitation threatens to degrade the resource base. For instance, in many Pacific island communities, marine resources are seemingly harvested under open-access conditions: there are few stated general rules limiting access, and if local residents are questioned about any such regulation they may say that all are free to fish as they like. However, as Hviding (1990) points out in a study of a Solomon Islands community, when resource extraction exceeds certain limits – commonly associated with the commercialization of fishing – marine tenure traditions begin to exert their force, and social sanctions limit the overexploitation by local residents of any particular area or species.[5] Similarly, the complex tenure and usufruct patterns of rainforest extractivists and shifting cultivators have recently been described (Colchester 1989). The existence of these invisible (to outsiders) or latent traditional management systems means that caution must be taken before judging any particular resource to be unregulated.

A third institutional mechanism increasing the sustainability of traditional resource use is the development, refinement and transmission of environmental knowledge in rural communities. Although often dismissed as 'intuitive', indigenous knowledge has in fact been distilled over centuries and is often the best guide to sustainable resource management. Perhaps the most striking and well-known example of a community which has a richly detailed knowledge of the plants, animals and soils of its environment, as well as of the best means of managing its resources in order to compensate for soil deficiencies, is the Kayapo of the Amazonian basin (Hecht 1989; Cummings 1990; Hecht and Cockburn 1990). Although the complexity of the ecosystem of the Amazon makes the Kayapo case particularly impressive, in fact, detailed indigenous environmental knowledge is the rule rather than the exception in traditional Third World societies (see e.g. Johannes 1981; Ravnborg 1990; Amanor 1990).

THE DYNAMICS OF THE COMMONS

Common property tenure and usufruct systems are central to many traditional management systems. Rural Third World communities often do not have as pervasive a sense of individual private property ownership as has developed in the industrialized world – instead, systems of group ownership prevail in many areas. It has, until fairly recently, been part of the conventional wisdom to believe that common property systems are inherently less productive, and more susceptible to degradation, than private property regimes. This belief was due to the metaphor of the 'tragedy of the commons' (generally attributed to Hardin 1968), which maintained that, because no individual would have to pay the full costs of over-exploitation, it would be in each individual's interest to extract as much as possible from the resource base, with the result that commonly held resources would inevitably be degraded. This view has largely lost theoretical support in recent years as the distinction between common property regimes (which consist in essence of jointly held property) and open access systems (which have no restrictions on resource use, and which are in fact subject to degradation) has become clear, and as more empirical studies have come out demonstrating the economic value of the commons (see e.g. Bromley and Cernea 1989).

Jodha's (1990) study has been particularly valuable in this latter regard. Research undertaken in 82 villages in India revealed that the poor obtain approximately one-fifth of their household income directly from common property resources, which in addition provide them with more than one-third of their farm inputs. Without the availability of common grazing lands in these communities, over half of the lands currently under food and cash crops would have to be diverted to fodder crops, or else livestock would have to be significantly cut, with a consequent drop in draft power and manure. In addition, Jodha argued, state interventions which have been undertaken to privatize common property, even when such interventions have been developed with the specific aim of helping the poor, have resulted in overall declines in the conditions of poor households.

However, in spite of the fact that the 'tragedy of the commons' scenario is no longer accepted by many development theorists, the metaphor remains a powerful influence on – or at least a strong basis for the rationalization of – the policies of both national governments and international development agencies which advocate settling pastoralists (Lane this volume; Adams 1990; Roe 1991), privatizing fishing grounds (Baines 1989; Polunin 1985), and supplanting traditional agricultural systems (Moorehead 1989; Diegues this volume). As Baines argues, 'there is a consistent tendency by agents of resource development to characterize traditional forms of resource administration as "problems" impeding development' (Baines 1989: 278). Thus as late as 1989 an official of the Tanzanian Ministry of Agriculture wrote, in terms directly recalling Hardin:

> [the] practice of grazing private livestock on communal land constitutes the single major constraint to improved management of the natural pasture lands. The inevitable result of this system of livestock production is that the cattle owners keep excessive numbers of livestock which in turn leads to overgrazing, soil degradation, low fertility and high mortality rates. ... Restriction of animal numbers to any reasonable balance with the forage resource has proved difficult due to lack of land ownership rights and communal land ownership.
>
> (quoted in Lane this volume: 99)

Similarly, the tragedy of the commons scenario has been explicitly built into the national rangeland policy in Botswana (Roe 1991).

However, efforts to avert the 'tragedy of the commons' themselves often become the cause of resource decline. For instance, when pastoralists are sedentarized, and restricted to the utilization of only part of the lands they have traditionally grazed, they are prevented from managing the remaining land in a sustainable manner. The restricted rangelands must now support many more livestock than would be permitted by traditional grazing regulations, and therefore often suffer degradation. It can thus appear that traditional pastoralism leads to overgrazing, when in fact it is the settling of pastoralist communities, and the disruption of their traditional common property management regimes, which triggers environmental decline in rangelands.

A key factor underlying the continued persuasive power of the tragedy of the commons metaphor is the belief that private land ownership gives individuals increased incentives for managing their resources sustainably. The argument is made that only people who have secure tenure over their landholdings will have the motivation to invest in the long-term undertakings necessary to ensure the continued yields of fragile environments. Indeed, several studies have shown the deleterious effects that lack of secure land tenure has had on local participation rates in environmental rehabilitation projects (e.g. Ståhl this volume). In most of these cases, however, lack of secure tenure is due not to the absence of private ownership as such, but rather to the fact that existing social structures allow those who control usufruct rights – whether they be individuals, groups, corporations or governments – to grant or withdraw these rights at will.

Common property management systems thus mistakenly get tarred with the same brush as some nationalized agricultural schemes, which, though in theory have the potential to be quite productive, in practice often suffer from unwieldy bureaucracies, insufficient resources, and the necessity of conforming to the demands of the international financial community. The actual or potential policy changes caused by these constraints create insecurity among the affected peasant farmers or pastoralists. Moreover, as Bromley and Cernea (1989) point out, the appearance that private property is more stable and adaptive than common property is due to the fact that the rights of exclusion for private property owners are generally upheld by the state: that is, the customary ability of private owners to exclude others from utilizing their land has been formalized and codified in law. On the other hand, the equally essential common property rights of exclusion, which have a firm basis and long history in common law, have been substantially eroded through the active or benign neglect of the state, and common property tenants are thus deprived of the legal protection afforded private owners.

In addition, it is clear that private ownership, secure land tenure, and sustainable resource use are not inevitably or intrinsically linked. For instance, small land owners who are obliged to go deeply in debt each season risk losing their land after a bad harvest; large land owners often show no qualms about clearing rainforests for short-term gains, even when it is clear that the resulting pasture lands will become barren in only a few years. Bandyopadhyay (1990) demonstrated that in certain communities in India, common property resources are better safeguarded than private property resources. The short time preferences of the private owners, and their ability to abandon degraded lands once maximum resources had been extracted, mean that they do not have the same incentives for environmental preservation that exist in communities whose families have inhabited a region for generations, and whose descendants will continue to inhabit it for generations to come. Kurien found the same phenomenon in his study of common fishing grounds:

> For the fishworkers, their future lies in the sea and its common resources. For capitalists, given their short-term perspective, and under the given conditions of investment, the ratio of profits from indiscriminate harvesting of the commons to the profits from regulated and sustainable harvesting are large. For them it actually pays to bring ruin to the commons!
> (Kurien this volume: 254–5)

THE SUSTAINABILITY OF TRADITIONAL SYSTEMS IN DEVELOPING SOCIETIES

As was discussed above, increased consumption is only one aspect of sustainable development. The question of traditional resource management, therefore, should not be examined only in terms of its efficiency in market economics: in many cases, traditional ways of interacting with the environment provide a fundamental basis for a community's well-being, and the abolition or support of traditional lifestyles thus becomes a human rights issue. When cultural and social identity is inextricably bound up in traditional forms of resource use, and when such communities desire to maintain this identity,[6] resource management policy decisions should not be made solely on the basis of which system will provide a maximum economic yield. Thus the plight of extractivist forest dwellers in the Amazon should not be dismissed because the marketing of their products must be subsidized, and the disappearance of the pastoralist way of life should not be considered an indispensable sacrifice to progress.

It is also important, however, not to idealize all indigenous practices or communal societies. Many traditional societies are clearly repressive, while even seemingly highly participatory traditional resource management systems can be inegalitarian, and common property can in reality exclude large numbers of people from enjoying the full benefits of its holdings. The exclusion of women from the decision-making and/or the benefits of such systems is perhaps the most readily observable example of inequality, although similar exclusions based on class, caste and race are also very common (Watson 1989).

To many in the development community, however, the question of the relative merits of different traditional systems is seen as moot: the common perception is that the sustainability of traditional ways of life is being threatened not only by exogenous pressures and policy decisions, but also by stresses coming from within the community, including increased integration into the market economy, increased contact with Western cultures, and population pressures. All of these factors do inevitably bring changes to lifestyles, but no tradition has ever been static, and change can occur without tradition being lost. The way in which current trends affect the sustainability of resource use in traditional societies, and thus the viability of such societies, remains a more open question, however. There are many examples of communities which lose their incentives to preserve the resources on which they no longer depend as capitalist development takes place, and they thus abandon their traditional management practices and eventually lose their knowledge of them. This process is not inevitable, however. Rainforest Indians can establish market relations with North American ice-cream chains without abandoning their relationship with the forest; African pastoralists can initiate political and economic contact with towns without losing their sense of reliance on the land; and many communities throughout the Third World have even opted to market their own 'indigenousness' to tourists, without giving up their own sense of the value of their way of life.

The relationship between population growth and resource degradation deserves special consideration because of the substantial attention it has received in recent years. The approach to this question has changed little since Malthus, and it is presently widely accepted that population growth will force resource extraction to exceed the capacity of the environment to renew itself, and environmental degradation will be the unavoidable result. Even among those who recognize that population pressure is not the only or ultimate cause of environmental problems (e.g. Shaw 1989), it is common to argue that reducing population growth is nevertheless the most effective means of arresting environmental decline.

In fact, however, evidence is mounting that the population growth approach is an oversimplistic means of portraying the environmental problems of the Third World. It is true, on the one hand, that in an ultimate sense the resources of the earth will be limited, and more specifically, that population pressures can contribute directly to overexploitation of resources in situations where people do not have available to them options which would allow them to adapt their behaviour in a sustainable way.[7] However, concentrating on slowing demographic growth rates in order to relieve particular environmental problems in the Third World can be ineffective and is at worst misguided, diverting attention from more fundamental causes and more productive solutions. As Somanathan (1991) demonstrates in a study of forest management in the Himalaya, deforestation in the region has historically been associated with government policy rather than changes in population size. When traditional forest management systems were disturbed in the 1920s, deforestation occurred in widening circles around villages in the span of a few years – clearly too short a time for a population explosion. The same study reveals that dense population does not necessarily imply deforestation: the crowded valley below the Chandag Reserve, which has retained its forestry control system, maintains well-protected *panchayat* forests, while the reserve itself, under government control, has been degraded. Similarly, deforestation in Brazil has been clearly demonstrated to have been influenced by a complex set of factors including policy decisions (Mahar 1989; Hecht and Cockburn 1990), although this does not prevent the continuing deforestation being ascribed to population pressure.

In addition, there is some evidence to show that the practice of overexploiting resources can be, under some circumstances, connected with an actual decline in population. A study undertaken in the Jebel Marra highlands of Sudan, for instance, describes a situation in which carefully managed agroforestry systems have been a part of the traditional environmental management practices of the region, and have helped to support a densely settled population for centuries (Miehe 1989). In recent years, however, the population has declined substantially (due in large part to the pull of newly accessible cities), and the resource management of the area has become less rigorous, with the result that tree cover has actually declined. The only mature tree plantations were planted over sixty years ago, and the knowledge that provided the basis for sound plantation management has now largely been lost. (See Blaikie 1985 for references to other cases in which population decline has led to environmental degradation.)

The need to refrain from overgeneralizing about the effects of population dynamics on traditionally sustainable resource management systems is well demonstrated by two studies of the traditional *milpa* agriculture of Mexico, a complex and highly developed form of resource management involving forest extraction, active fallow management and cultivation of maize and other crops. Barrera Bassols *et al.* (forthcoming) report on an indigenous community in northern Veracruz, detailing the traditional agricultural practices which have remained productive for generations in the context of a strong tradition of community identity, shared labour and a conscious effort to ensure the integrity of the environment. Recently, however, population growth seems to have reached the point at which the traditional low-input, shifting cultivation techniques will no longer be feasible: in 1989 chemical fertilizers were used for the first time. García Barrios and García Barrios (forthcoming) report on changes taking place within the *milpa* of another community in Oaxaca. In this case, dramatic declines in population due to outmigration have resulted in a situation similar to the Jebel Marra case: the residents are beginning to lose the ecological knowledge which formed the basis of the *milpa* system, and new techniques are not being developed to replace it.

Although the outcome of these two Mexican cases is in some ways similar, in that the traditional *milpa* agriculture has come under pressure from endogenous forces, the indications are that the eventual outcome will be quite different for the two communities. In the Oaxaca case, agricultural production is declining faster than are the needs of the community, which are being reduced through outmigration, while in the Veracruz case, the growing community has maintained not only food self-sufficiency, but also substantial agricultural surpluses. The eventual outcome of the changes taking place in the Veracruz community remains to be seen, but it is quite possible that a new form of 'traditional' resource management will be developed which will continue to enable this community to fulfil its needs. From these two examples, it is clear that population growth (and indeed population decline) is only linked to unsustainable resource use to the extent that the population in question does not have the means to adapt its resource management practices to the changing needs of the community.

In summary, the evidence shows that making generalized judgements about the future of traditional resource management systems in the context of development is inappropriate. It cannot be said that it is either possible or desirable to maintain all such systems, and at the same time it is a mistake to dismiss them as obsolete or unable to remain adaptable in the face of endogenous pressures.

What is clear, however, is that the presence of traditionally sustainable environmental practices can provide opportunities for achieving sustainable development which should not be overlooked. In some cases existing systems can arguably be maintained, at least for the medium-term future, in the absence of interference (see e.g. Polunin 1985; Lane this volume; Cummings 1990). In other cases, traditional resource management techniques show potential for use in informing successful new resource management initiatives (see e.g. Draz 1985; Bromley and Cernea 1989; Yabes this volume). A third possibility is that traditions of community management of resources will form the basis of community action which specifically addresses environmental issues. Diegues (this volume) argues that the presence of traditional communities can be considered insurance that the environment will be conserved, provided that their management schemes remain viable, because such communities will not allow environmental degradation if it is in their power to arrest it. It is this potential of traditional systems to provide the foundation of popular initiatives that is the subject of the following section.

TRADITIONAL SYSTEMS AND POPULAR INITIATIVES

In the face of the internationalization of even very local economies, increasing commercialization, and direct pressure and hostility from development agents, it cannot be assumed that traditional resource management systems can continue as before without the active support and struggle of their participants. Such struggles, based on the efforts of local people to maintain or improve their levels of living by halting resource degradation which threatens their traditional livelihoods without providing new benefits to them, are occurring in many parts of the Third World. Perhaps the best-known are the Chipko movement – which originated with localized efforts to prevent the destruction of Indian forests by loggers, and has developed into a regional movement with wide-ranging environmental concerns – and the rubber tappers' association – which was organized to protect the Brazilian forest-dwellers' rights to extract forest products in a sustainable way, and has resulted in the establishment of extractivist reserves protected from logging (Bandyopadhyay and Shiva 1988; Schwartzman 1989). These two movements have become subjects of much discussion, but their frequent citation should not obscure the fact that such struggles are in fact quite widespread.

Reaction against infrastructure projects, such as dams or roads, which threaten a transformation or an outright destruction of the environment is perhaps the most common form of what may be termed environmental resistance movements. Again, it is the large projects – the Narmada dam in India or the Balbina in Brazil – which have attracted the most attention, but these are far from being the only examples of such activities. The Jonglei canal project in southern Sudan, which has been planned since the 1940s as a way to bypass the swamps of the region, is as yet incomplete in part because local resistance (influenced by the civil war) enabled serious environmental flaws in the design to be recognized. In Nigeria, resistance to the Bakolori reservoir project in the late 1980s was violently repressed (Adams 1990), but the movement ultimately had repercussions elsewhere in western Africa, as later dam projects made more deliberate efforts to address the concerns of local communities.

Environmental activism does not only occur when complete environmental destruction is threatened, but can also result from attempts to convert resources from one form to another in a way which renders traditional ways of life untenable, without providing alternative economic opportunities to the communities affected. Lane's work with the Barabaig pastoralists in Tanzania (this volume) gives one example of the form which such activism can take. The conversion of part of their traditional grazing lands to wheat farms forced the Barabaig to concentrate their herds on the inferior lands remaining to them. As a consequence, both this land and the area converted to wheat monocropping have suffered soil erosion. Economic hardship for the Barabaig has been the result, while the wheat farms have failed to yield the benefits – to the farmers or to the country – which were predicted. The Barabaig have reacted by challenging in court the process which resulted in the appropriation of the land, and asserting their claim that their customary use of the land should grant them legal title to it. The case has not yet been decided, but it has attracted worldwide attention, not least from the Canadian funders of the wheat farm project. Even if the Barabaig lose their case, their activity will have resulted in closer examination by at least some donors of schemes which transform rangelands to farms.

Similarly, the conversion of Brazilian wetlands, traditionally managed by fishing and farming communities, to irrigated sugar cane and rice plantations has resulted in collective action. In his study of *várzea* (floodplain) fishing communities in Brazil, Diegues (this volume) found that without this social mobilization, neither the livelihood of traditional communities nor the environment can be conserved, because the plantations not only displace the traditional inhabitants of the area, but also cause significant environmental degradation, disturbing the finely tuned ecosystem which maintains marine life. As in the Tanzanian case, the conversion to farming of parts of former communal land affects surrounding areas as well. Not only have certain species of fish and trees disappeared altogether, but the remaining fishing areas must be exploited more intensively than before. Protests against the plantation schemes began in 1986, and enlisted the assistance of national and international support groups. The ecological and cultural importance of the *várzea* was documented in a series of technical surveys, and by 1988 the area was declared under environmental protection.

Environmental activism in the Third World involves not only struggle against the expropriation of resources, but also resistance to resource overexploitation by outsiders. The communities of the rainforests of Sarawak, in Malaysia, whose economy and way of life are based on the forest, have protested against government-supported logging activities whose benefits are channelled to elites outside of the region. In 1987 the Penan, a hunting and gathering society appealed to the government to stop the logging:

> Stop destroying the forest or we will be forced to protect it. We have lived here before any of you outsiders came. We fished in clean rivers and hunted in the jungle.... Our life was not easy, but we lived it in content. Now, the logging companies turn rivers into muddy streams and the jungle into devastation. The fish cannot survive in dirty rivers and wild animals will not live in devastated forest.... We want our ancestral land, the land we live off, back. We can use it in a wiser way.
>
> (quoted in Colchester 1989: 42)

The logging did not stop, and the Penan people acted, blockading logging roads. The action was soon taken up by other communities of the rainforest, and logging activity was brought to a standstill. In addition, the activists were able to link into environmental networks to publicize their cause, and to launch an international campaign to influence other countries to stop importing Malaysian timber. Although some concessions have been won, the campaign has not been an unqualified success: 'crusading' Western environmentalists were accused of exploiting the Penan and hindering development, while at the same time Penan leaders were jailed under an anti-terrorist law (Lim 1989; *Utusan Konsumer* 1989). More recent action to stop logging has been met with similar measures (*Far Eastern Economic Review* 1991).

The Malaysian experience has been repeated in many areas of the world where forest communities have been threatened by logging, including India, Thailand, the Philippines and Brazil. Activism in response to overexploitation of common resources by outsiders is not limited to indigenous forest communities, however. Kurien's study of the responses of the fishworkers of the Kerala coast to overfishing of common waters by commercial fishing companies is a case in point (this volume). As it became clear that trawler fishing was depleting fish stocks to an unsustainable degree, a campaign to ban the large boats from the coastal waters during the crucial spawning season was begun. It took almost ten years of organization, agitation, hunger strikes and political manoeuvring, but by 1988 such a ban was enacted.

Again, it cannot be said that success was complete: the struggle had to be continued in order for the fishing ban to be imposed the following year. At the same time, the years of competition with the commercial fleets had caused many of the traditional fishermen to turn to outboard motors and miniature versions of the destructive ring seines used by the large boats in order to increase their own catches; now the traditional fishing community is in danger of abandoning its previous sustainable practices. However, this trend is offset by another one. In the course of activating the collective action against the commercial boats, many of the organizers have gained a more precise understanding of the limits of marine resources, and they are now beginning to act to stop the newly formed destructive habits of their own community.

THE SIGNIFICANCE AND POTENTIAL OF LOCALLY BASED ENVIRONMENTAL INITIATIVES

There are several lessons to be learned from a study of environmental activism undertaken by traditional communities. It can be observed that collective action to resist the implementation of environmentally destructive development projects is rarely triggered primarily by an overriding concern to preserve the environment in its existing state, but rather hinges on the lack of sufficient benefits from such projects accruing to local communities. This fact does not imply that traditional communities are insensitive to the aesthetic niceties of their surroundings, but rather indicates that they have a desire to survive, and to improve their living levels and consumption levels if possible.

It has also become clear that, because of their extensive ecological knowledge, societies which are based on sustainable environmental management practices are much better able to accurately assess the true costs and benefits of ecosystem disturbance than any evaluator coming from outside the local area. Such societies are the first to realize that 'development' which results in environmental degradation will rarely yield net benefits in the long run, and the emergence of popular opposition to a project can therefore be taken as an important warning sign that it will have negative environmental consequences.

A third indication of the investigation into the dynamics of environmental activism is that the success of such movements is often due to their ability to form a coalition with regional, national or international groups which have similar interests, and to publicize their grievances and their cause. Such support for local-level activity can come from NGOs with development and equity concerns, from social movements focusing on human rights issues, or from international agencies directed towards environmental conservation. In addition, it appears that among the residual benefits of collective action of this sort is the ability of many of the movements to turn from negative to positive activity – they move, that is, from opposing to initiating activities in a process that Hirschman has dubbed 'the conservation and mutation of social energy' (Hirschman 1984). This transformation has taken many forms – from opposition to a large dam resulting in proposals and support for a series of smaller, more manageable dams (Bandyopadhyay 1990), to the formulation of the entirely new concept of extractivist reserves (Schwartzman 1989).

It is also clear, however, that environmental activism does not take place in a vacuum. The impact of such movements – and, indeed, the possibility of collective action being undertaken at all – depends to a large extent on the social, economic and political structures which influence community dynamics from the local, national and international levels. Thus the Kerala fishworkers' movement owes a good part of its relative success to the fact that the fishworkers were working within a social and political system which enabled them to form a voting block large enough to make themselves felt at the state government level, in spite of the superior resources of the opposing fishing lobbies. The Penan, on the other hand, have had a substantially smaller real impact despite their more intensive and desperate struggle: the mechanisms by which the needs of these forest dwellers could come to outweigh the powerful logging interests are much weaker in this case, and in fact the progress which has been made is due in large part to international pressure, rather than to government responsiveness to its weaker constituents.[8] The importance of structural factors for the success, or even the existence, of collective action means that such action is not undertaken in all circumstances where there is a need and a will to do so. A repressive state can crush or atomize organizational efforts at an early stage, while the domination of the economy by outside interests can close off channels of activity on the local level. Similarly, the existence of intra-community repression can prevent class, race or gender-based alliances from forming, and from making their interests felt.

A fifth indication of this research is that the need for activism around local environmental issues has put sustainable resource management on the agenda of activist groups and NGOs with wider concerns. Thus the Organisation of Rural Associations for Progress (ORAP), a Zimbabwean development NGO, responded to the environmental degradation caused by inappropriate green revolution farming techniques by supporting a return to drought-resistant crops, organic farming methods and afforestation programmes (Nyoni 1990). A similar process has taken place within other popular organizations (see e.g. Blauert and Guidi this volume). For example, the Popular Defence Committee of Durango, Mexico, which was originally organized to obtain housing rights and other basic needs, turned to environmental activism when industrial pollution threatened the water supply of the community. In time, this socially based ecological movement not only widened to include surrounding rural areas in its activities, but also expanded its activities to address problems of sewage disposal, drainage and refuse management (Moguel and Velázquez this volume).

CONCLUSION: POVERTY, EMPOWERMENT AND SUSTAINABLE DEVELOPMENT

There is at present much discussion of the relationship between poverty and environmental degradation. The common argument is that poor people are forced to cultivate marginal lands, or to overexploit resources in spite of the fact that they threaten their future livelihood by doing so, because they will not otherwise be able to survive the present season. There is some truth in this model, but as a basis for understanding the primary causes of environmental degradation it is not particularly helpful. It may just as easily be said (as it has often been) that the excessive wealth and overconsumption of industrialized societies is responsible for the vast majority of unsustainable resource extraction, and that wealth may therefore be more appropriately blamed for ecological problems than poverty. Again, however, while this way of framing the problem may be accurate, it does not provide substantial assistance in finding ways out of the difficulties. And, as the scale of the ecological disasters in some of the former socialist countries attests, neither does the presence of relative economic inequality fully account for environmental degradation.

The research on participation in resource management described above, however, has shown that poor communities not only have high incentives for managing their resources sustainably, but they have historically often been able to develop a variety of effective and adaptable means of doing so. Environmental degradation in rural areas of the Third World is not due to the poverty of rural communities; rather, poverty is a symptom of one of the primary underlying causes of local-level environmental decline in the Third World today: the disempowerment of these communities.[9] Furthermore, to the extent that local-level degradation forms a major component of global environmental problems, this growing inability of communities to participate in resource management decisions has an important impact on the potential for sustainable development. Disempowerment in the course of development can take many forms. People may be deprived of access to the resources on which they depend, their traditional tenure rights and rights to exclude outsiders may be abrogated, or their ability to make their own decisions regarding resource management may be curtailed. In all of these cases the result is similar: resource management decisions are taken over by those with insufficient stake in the local environment, and resources are extracted at unreplenishable levels in order to benefit other, often richer, societies.

It is clear, then, that struggles for greater participation are essential elements of the foundation of an endurable basis for sustainable development. This process can only be helped by the growing recognition of the importance of the environment for the future well-being of the entire planet. Muntemba argues that 'environmental degradation is becoming a liberating force':

> I have come across communities/chiefdoms where the eco-
> logical degradation which found expression in the food crisis
> of the 1980s pushed people into taking conservation measures
> which in fact flouted national laws.... Some governments are
> willing and anxious to try ways of managing the natural
> resources to ensure the livelihoods of poor people. Where
> people know this, they are seizing the opportunity for further
> empowerment.
>
> (Muntemba 1990: 4)

Under certain conditions it is clear she is right – in particular when degradation reaches the point where outside interventions are abandoned in favour of local initiatives, or when an ecological disaster resulting from interventions in one area leads to calls for more recognition of the sustainability of traditional practices in similar ecological zones. The question of how widespread a phenomenon this empowerment process can become, however, is still open. It will depend on the efforts of both development agents and environmentalists not only to support people's rights for self-determination, but also to recognize that their struggles are essential for the health of the environment.

NOTES

1 This chapter has benefited greatly from comments by Solon Barraclough, Dharam Ghai and Cynthia Hewitt de Alcántara, although remaining errors and omissions are my own.
2 See the introductory chapter for further discussion of how the term 'participation' is understood in this volume.
3 This is implicitly understood in some uses of the term sustainable development: at times the North seems to see itself as entitled to take measures to enforce environmental preservation without regard to the needs of the South – as when, for instance, Northern writers lament 'our vanishing rainforests'.

4 Although there remains, as well, a strong anti-growth trend of thought among more fundamentalist ecologists.

5 More active sanctions have been used to combat commercial intrusions from outsiders, ranging from sabotage and assault to the imprisonment of a US fishing boat captain and the impoundment of his vessel in the Solomon Islands in the early 1980s.

6 These, of course, are immensely complex issues, which, not being central to the argument of this chapter, will not be further discussed here.

7 The importance that the power of adaptation has for the outcome of population growth is demonstrated by the fact that, as the Brundtland report points out, 'a child born in a country where levels of material and energy use are high places a greater burden on the Earth's resources than a child born in a poorer country' (WCED 1987: par. 4.48). The emphasis of population programmes remains on the South, however, while many Northern countries are undertaking concerted efforts to increase their own birthrates.

8 At least one Malaysian activist argues that the solution to this problem is not to change governments, but to break the power of multinational corporations to dictate government policy on issues such as these.

9 A full discussion of the concepts of 'empowerment' and 'disempowerment', and of the problems with these terms, is beyond the scope of this chapter. As used here, 'empowerment' refers to a complex process centred around people's efforts to increase their participation – that is, their control over physical and social resources – within the development process.

REFERENCES

Adams, W.M. (1990) *Green Development: environment and sustainability in the Third World*, London: Routledge.

Amanor, Kojo (1990) 'Land degradation, fallow management and local environmental knowledge in the Krobo district of Ghana', Geneva: UNRISD, mimeo.

Baines, G.B.K. (1989) 'Traditional resource management in the Melanesian South Pacific: a development dilemma', in Berkes (ed.).

Bandyopadhyay, J. (1990) 'From natural resource conflicts to sustainable development in the Himalaya', paper prepared for UNRISD workshop on 'Sustainable development through people's participation in resource management', 9–11 May.

Bandyopadhyay, J. and Shiva, V. (1988) 'Political economy of ecology movements', *Economic and Political Weekly*, 23 (24): 1223–32.

Barraclough, Solon (1990) 'Popular participation in rural development and forestry', paper presented at the regional workshop on 'Implementation of participatory forestry projects', *Debre Zeit*, October.

Barrera Bassols, Narciso, Ortíz Espejel, B. and Medellín, S. (forthcoming) 'Un reducto de abundancia: el caso excepcional de la milpa en Plan de Hidalgo, Veracruz', in Hewitt de Alcántara (ed.) forthcoming.

Berkes, Fikret (ed.) (1989) *Common Property Resources: ecology and community-based sustainable development*, London: Belhaven Press.

Blaikie, Piers (1985) *The Political Economy of Soil Erosion in Developing Countries*, Harlow, Essex: Longman Scientific and Technical.

Bromley, Daniel W. and Cernea, Michael M. (1989) 'The management of common property resources: some conceptual and operational fallacies', Washington, DC: World Bank, discussion paper 57.

Browder, John O. (ed.) (1989) *Fragile Lands of Latin America: strategies for sustainable development*, Boulder, Co.: Westview Press.

Colchester, Marcus (1989) *Pirates, Squatters and Poachers: the political ecology of dispossession of the native peoples of Sarawak*, London: Survival International.

Cummings, Barbara J. (1990) *Dam the Rivers, Damn the People: development and resistance in Amazonian Brazil*, London: Earthscan.

Draz, Omar (1985) 'The Hema system of range reserves in the Arabian Peninsula: its possibilities in range improvement and conservation projects in the Near East', in McNeely and Pitt (eds).

Far Eastern Economic Review (1991) 'Anti-logging protesters held in Sarawak', 18 July, p. 14.

Farvar, M. Taghi (1987) 'Local strategies for sustainable development', in Jacobs and Munro (eds).

Gadgil, Madhav (1985) 'Social restraints on resource utilization: the Indian experience', in McNeely and Pitt (eds).

García Barrios, Raúl and García Barrios, Luis (forthcoming) 'Subsistencia maicera y dependencia monetaria en el agro semiproletarizado: una comunidad rural Mixteca', in Hewitt de Alcántara (ed.).

Hardin, G. (1968) 'The tragedy of the commons', *Science*, 162: 1243–8.

Hawkins, Ann P. and Buttel, Frederick H. (1990) 'The political economy of "Sustainable Development"', Ithaca, New York: Cornell University, Program on Science, Technology and Society, mimeo.

Hecht, Susanna (1989) 'Indigenous soil management in the Amazon Basin: some implications for development', in Browder (ed.).

Hecht, Susanna and Cockburn, Alexander (1990) *The Fate of the Forest: developers, destroyers and defenders of the Amazon*, London: Penguin.

Hewitt, de Alcántara Cynthia (ed.) (forthcoming) *Restructuración económica y Subsistencia Rural: El Maíz y la Crisis de los 80*, Geneva: UNRISD.

Hirschman, Albert O. (1984) *Getting Ahead Collectively: grassroots experiences in Latin America*, New York: Pergamon Press.

Hviding, Edvard (1990) 'Keeping the Sea: aspects of marine tenure in Morovo Lagoon, Solomon Islands', in Kenneth Ruddle and R.E. Johannes (eds), *Traditional Marine Resource Management in the Pacific Basin: an anthology*, Jakarta, Indonesia: UNESCO/ROSTSEA.

IUCN (International Union for the Conservation of Nature and Natural Resources) (1980) *World Conservation Strategy: living resource conservation for sustainable development*, Gland, Switzerland: IUCN/ UNEP/WWF.

Jacobs, P. and Munro, D.A. (eds) (1987) *Conservation with Equity: strategies for sustainable development*, Gland, Switzerland: IUCN.

Jodha, N.S. (1990) 'Rural common property resources: contributions and crisis', *Economic and Political Weekly*, 30 June, A65–A78.

Johannes, R.E. (1981) *Words of the Lagoon: fishing and marine lore in the Palau District of Micronesia*, Los Angeles: University of California Press.

Lim Siong Hoon (1989) 'Malaysia in skirmishes over environment', *Financial Times*, 27 October.

Mahar, Dennis J. (1989) *Government Policies and Deforestation in Brazil's Amazon Region*, Washington, DC: World Bank.

McNeely, Jeffrey A. and Pitt, David (eds) (1985) *Culture and Conservation: the human dimension in environmental planning*, London: Croom Helm.

Miehe, S. (1989) 'Acacia albida and other multipurpose trees on the farmlands in the Jebel Marra Highlands, Western Darfur, Sudan', in Nair (ed.).

Moorehead, Richard (1989) 'Changes taking place in common-property resource management in the inland Niger delta of Mali', in Berkes (ed.).

Muntemba, S. (1990) 'People's participation in environmentally sustainable development: a reflection on some emerging issues', paper prepared for UNRISD workshop on 'Sustainable development through people's participation in resource management', 9–11 May.

Nair, P.K.R. (ed.) (1989) *Agroforestry Systems in the Tropics*, Dordrecht: Kluwer Academic Publishers and ICRAF.

Nyoni, Sithembiso (1987) 'Indigenous NGOs: liberation, self-reliance, and development', *World Development*, vol. 15 Supplement (Autumn), 51–6.

Nyoni, Sithembiso (1990) 'Sustainable development through people's participation in resource management: the case of ORAP, Zimbabwe', paper prepared for UNRISD workshop on 'Sustainable development through people's participation in resource management', 9–11 May.

Pearse, Andrew and Stiefel, Matthias (1979) 'Inquiry into participation: a research approach', Geneva: UNRISD.

Polunin, Nicholas V.C. (1985) 'Traditional marine practices in Indonesia and their bearing on conservation', in McNeely and Pitt (eds).

Ravnborg, Helle Munk (1990) 'Peasants' production systems and their knowledge of soil fertility and its maintenance: the case of Iringa district, Tanzania', Copenhagen: Centre for Development Research, working paper 90.1.

Rees, William E. (1990) 'The ecology of sustainable development', *The Ecologist*, 20 (1) (January/February): 18–23.

Roe, Emery M. (1991) 'Development narratives, or making the best of blueprint development', *World Development*, 19 (4): 287–300.

Sankhala, K.S. and Jackson, Peter (1985) 'People, trees and antelopes in the Indian desert', in McNeely and Pitt (eds).

Schramm, Gunter and Warford, Jeremy J. (eds) (1989) *Environmental Management and Economic Development*, Washington, DC: World Bank.

Schwartzman, Stephan (1989) 'Extractive reserves: the rubber tappers' strategy of sustainable use of the Amazon rainforest', in Browder (ed.).

Shaw, Paul (1989) 'Rapid population growth and environmental degradation: ultimate versus proximate factors', *Environmental Conservation*, 16 (3): 199–208.

Somanathan, E. (1991) 'Deforestation, property rights and incentives in Central Himalaya', *Economic and Political Weekly*, 26 (4): PE 37–PE 46.

Thrupp, Lori-Ann (1989) 'Politics of the sustainable development crusade: from elite protectionism to social justice in Third World resource issues', University of California Berkeley, Energy and Resources Group, mimeo.

Tobayiwa, Chris (1985) 'Shona people, totems and wildlife', in McNeely and Pitt (eds).

Utusan Konsumer (1989) 'Penan plight moves the world', Penang, Malaysia: Consumers' Association of Penang, No. 195, November.

Warford, Jeremy (1987) 'Environmental management and economic policy in developing countries', in Schramm and Warford (eds).

Watson, Dwight J. (1989) 'The evolution of appropriate resource-management systems', in Berkes (ed.).

WCED (World Commission on Environment and Development) (1987) *Our Common Future*, Oxford: Oxford University Press.

Part II

TRADITIONAL SYSTEMS OF RESOURCE MANAGEMENT

4

THE BARABAIG PASTORALISTS OF TANZANIA: SUSTAINABLE LAND USE IN JEOPARDY

Charles Lane

INTRODUCTION

The failings of the green revolution have shattered faith in the power of technology alone to solve development problems. Development has come to mean more than simply increasing production. Wider considerations of social equity and environmental conservation have proved to be equally important. Economists are also now realizing that the pursuit of economic growth has an environmental down side that must be given a cost in future budgeting (Pearce *et al.* 1989). Conservationists from all over the world are involving themselves in the struggle to win the battle to protect Africa's natural resources (Harrison 1987). Social scientists are also now revealing the value and importance of traditional land use systems for the preservation of people's livelihoods and the conservation of nature. Sustainable development has become a new and important goal for developers (Conway and Barbier 1988).

In recent years more attention has been given to pastoralists' traditional resource management systems, particularly those with common land tenure arrangements (see Baxter 1989; Raintree 1987; National Research Council 1986). From this has come a growing body of opinion that these systems are both economically efficient and sustainable (Abel and Blaikie 1989). However, this is not yet adequately reflected in pastoralists' involvement in policy formulation or development performance.

Tanzania is blessed with some of the richest pastoral resources in Africa. Foremost amongst these are pastoralists and their traditional

81

natural resource management systems. To date most attempts at pastoral development in East Africa have mostly failed for lack of understanding and adherence to local land tenure arrangements (Lane and Swift 1989). A development project in a pastoral area of Tanzania offers a case that highlights the costs of failing to recognize and support traditional pastoralism. The Tanzanian government, with Canadian aid, has failed to take account of new advances in development thinking and has persisted with a project that is environmentally destructive, socially unjust and economically unsound.

Barabaig pastoralists of the Hanang plains, Hanang district, Arusha region, northern Tanzania, are currently enduring the negative impacts that result from having their pastures taken from them for a government wheat scheme. This is denying them the means for production, causing land degradation, and impoverishing them. It is therefore timely to examine what is happening to the Barabaig and see whether their land use system might offer a sustainable alternative to this project.

By providing details of the traditional Barabaig pastoral land use system, it will be argued that the way the Barabaig manage natural rangeland resources is a rational and sustainable form of land use. Evidence will be given to show that their common land tenure arrangements are both sophisticated and effective for production and conservation of land. However, an inappropriate and costly development project has undermined this system, causing the Barabaig to suffer and destroying rangeland. By failing to accommodate Barabaig land use arrangements and support Barabaig pastoral production, this project is ignoring the means by which natural rangeland resources could have otherwise been conserved.

THE BARABAIG OF THE HANANG PLAINS

The Barabaig are semi-nomadic Nilotic pastoralists who are a sub-tribe or section of a wider ethnic grouping called Tatoga.[1] They number between 30,000 and 50,000 in Hanang district.[2] They fled from an invasion of the more numerous and powerful Maasai and surrendered occupation of the Serengeti and Ngorongoro highlands more than 150 years ago (Saitoti 1986). They moved south along the Rift valley as far as Singida before returning to what they now regard as their territory on the Hanang plains (Wilson 1952).

All Barabaig herders strive to be self-sufficient from production of their cattle. Each household head manages his herd to maximize production of milk, meat and occasionally blood. However, they do not exist on a purely pastoral diet. Only those with large herds receive half or more of their food needs from cattle products. Maize makes an important contribution to the nutrition of all Barabaig, but most especially poorer households with fewer cattle. Grain is obtained through exchange or sale of livestock, and from shifting cultivation by households with the help of communal labour provided by relatives and neighbours.

The Hanang plains are divided by the Great Rift Valley escarpment which separates the northern and elevated Basotu plains from the lowland Barabaig plains to the south (see Figure 4.1). The Basotu plains are undulating with a series of depressions and many low hills and volcanic craters. As there are no perennial rivers, crater lakes provide some of the few sources of permanent water. However, even with the provision of dams, built in colonial times, there is a severe shortage of water in the dry season. The Barabaig plains,[3] on the other hand, have permanent water from lakes Balangda Gidaghan and Balangda Lelu.[4] These lakes also provided salt for the Barabaig and their livestock. Mount Hanang dominates the landscape rising to a peak of 3,118 metres above the surrounding Barabaig plains. The plains extend to the south as far as the Mureru range. Apart from bushland beyond the range, and the forest of Mount Hanang, the dominant vegetation is acacia and commiphora woodland interspersed with open grassland. The climate is semi-arid with periodic droughts and an average annual rainfall of around 600 millimetres.

PEOPLE, CULTURE AND LAND

People, culture and land are inseparable elements of human ecology. As such, they are a key to sustainable development. This is as true for people of rich industrial nations as for those of the developing world. Traditional cultures, however, have better maintained the relationship between these three elements. Some peoples have existed on the same pasture resources held in common for hundreds of years without destroying them (Netting 1978). One of the reasons for their success arises from the way they regard land. They accept that they are an integral part of a wider whole that includes soils, water, vegetation and animals. Being members of a larger entity, they treat land with the utmost respect. They feel able to

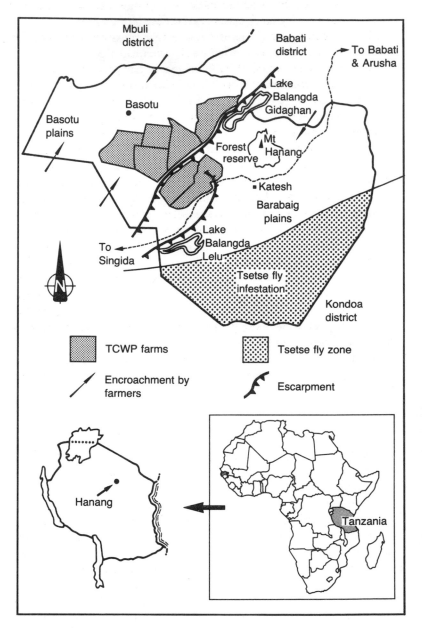

Figure 4.1 Hanang district
Source: Lane 1991a

84

enjoy its bounty, but not regard it as a commodity and exploit it without considering its future preservation.

The Barabaig regard land as more than a physical resource. For them it can also have great spiritual significance. The Barabaig bury certain of their esteemed elders with a *bung'ed*, which is both the name of the burial mound and the funeral ceremony associated with it. The ceremony involves thousands of people and costs what amounts to a fortune for the Barabaig, including as it does the slaughter of many livestock and the brewing of vast quantities of honey beer. They hold celebrations at the grave site over a period of nine months. An earthen mound (three generations ago the mounds were made of stones) is built up on the grave that eventually rises to a height of three metres. The ceremony culminates when the eldest son of the deceased's first wife climbs the mound and offers a prayer. Before he descends he places his father's stick and sandals on top. Thereafter, the deceased man's clan is forever responsible for the grave's upkeep. Clansmen will visit it for generations to appeal to their ancestor as a medium to *Aset*, their God. The Barabaig still visit the *bung'ed* of Gitangda in the Ngorongoro crater that remains from the time they occupied the area before they were dislodged by the Maasai over a hundred years ago (Borgerhoff-Mulder *et al.* 1989). These burials constitute one of the most important expressions of Barabaig culture. The presence of a *bung'ed* provides an attachment to land that cannot be broken in Barabaig culture.

Traditional seasonal grazing rotation

Consistent with the pastoral ideal, the Barabaig make every effort to serve the needs of their livestock. Location of habitation and movement of Barabaig households, therefore, is primarily determined by considerations of animal husbandry.

Variety of soil types, topography, vegetation and the availability of ground water provide the basis on which the Barabaig classify pastures. They recognize specific forage regimes that are associated with eight geographical features. So as to utilize pasture when it is most productive, and rest areas and allow them to recover, the Barabaig have devised a seasonal grazing rotation that exploits the forage regimes at different times of the year. This is particularly so in drought years when variable use of pastures is even more important to make the most of scarce resources. The rotation is not

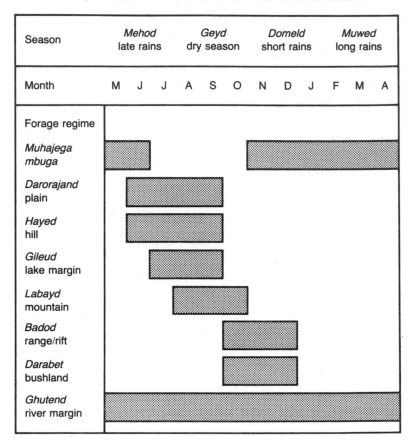

Season	Mehod late rains		Geyd dry season		Domeld short rains		Muwed long rains	
Month	M J J	A S	O N	D J	F M A

Figure 4.2 Traditional grazing rotation
Source: Lane 1991a

always regular as climatic variation can cause deviations (Figure 4.2). Presented here is a stylized model of a complex and variable cycle.[5]

The Barabaig year begins in May. This is the start of the late rains (*mehod*) when pasture availability and livestock production are at maximum. It is a time of relative plenty and recovery from the deprivations of the dry season. It is a time of celebration when a surplus of production and freedom from labour allows people to perform ceremonies that would otherwise be beyond their capacity to organize and supply.

It is also the time when the Barabaig like their herds to be on the *muhajega* (singular *muhed*). These are depressions on the plains containing fertile soils that sustain a mix of grasses and herbs that the Barabaig call *nyega nyatk*. The Barabaig regard the forage of the *muhajega* as the most productive pasture available to them. It is valued for its capacity to produce high milk yields, growth inducing capacity and recuperative powers for livestock suffering ill health from the stresses of the dry season and droughts. The *muhajega* are found in greatest abundance on the Basotu plains. It is on these plains that the Barabaig congregate during the rains. This is only possible for as long as surface water is available.

When the rains stop in May, June or July, and surface water dries up on the Basotu plains, many Barabaig move down the rift escarpment onto the Barabaig plains. Here they are able to draw on permanent water from wells at Lake Balangda Lelu that sustain them through to the months of August, September and October of the dry season (*geyd*). At this time they gather on the plains (*darorajega*) of Balangda and the many low hills (*hayed*) to the east of the lake. As grazing becomes depleted, they begin to exploit the lake margin itself (*gileud*) and move to the mountain slopes (*labayd*). As dry season grazing becomes increasingly scarce, herds are moved further afield to the rift escarpment in the north and the Mureru range (*badod*) to the south. This is also the time when they are forced to enter bushland (*darabet*) with its associated risks from tsetse flies and predation from wild animals.

As soon as the short rains (*domeld*) begin in November or December, and surface water is found again on the Basotu plains, the Barabaig move their herds back onto the *muhajega*. They remain there for the six or more months of the rainy season (*domeld*, *muwed* and *mehod*) until the water dries up again. But for the river margin (*ghutend*), the *muhajega* supports the Barabaig for longer than any other forage regime. As the *ghutend* is confined to the few small perennial rivers that flow a short distance from Mount Hanang, it is much less important than the *muhajega*, which the Barabaig value more than any other forage regime.

Thus, for the Barabaig, availability of water acts as the most limiting factor in pasture use. Whereas most other pastoralists need to locate pastures in the dry season to sustain them (sometimes they are preserved for this purpose), the Barabaig are forced to move away from their richest forage regime when they need it most. This means that they have to preserve the less productive *darorajand* for the dry season when they need to congregate there and be near the permanent water afforded by wells at Lake Balangda Lelu.

Common property resource management

Barabaig movement of herds and homesteads around the plains is a response to the vagaries of climate, and so as to make best use of scarce and variable pasture resources. To make this possible, the Barabaig accept that everyone must have general access to pastures. Open rangeland is therefore regarded as the property of the whole community. This enables herders to choose when and where to be on the plains according to their individual assessment of pasture quality.[6] So as to facilitate this peripatetic use of forage resources, the Barabaig, like other African pastoralists, have a form of common land tenure (see for example Makec 1988). They describe the land that makes up the range as *ng'yanyida madagh* (*madagh* = common). However, this is not a case of universal and uncontrolled access to land and its resources by everyone and anyone.

The Barabaig recognize the right to protect land: *weta ng'yanyid* (*weta* = protect). In the past, the Barabaig provided protection against intruders of their territory by deployment of the warrior set. Due to the breakdown of the process that formed this socio-political unit, and as a result of the imposition of central law and order by the state, they no longer are able to defend their territory as they had done before.

All Barabaig accept that their use of land is limited to the right of *usufruct* which permits use of common land only to the point where it is not denuded beyond recovery or to the disadvantage of other users. Overuse is regarded as destruction of the means on which everyone depends.

Land is made up of a bundle of rights that apply to a number of disaggregated resources. These can include certain vegetation species, ground water or trees. Rights of use to some land can be enjoyed by the whole community but certain limits do apply to resources within it.

Within the commons, rights to property range from communal access to exclusive private property. The Barabaig regulate rights of use and access to land through a tripartite jural structure, each with its own sphere of interest and authority; the community, the clan and individual households.[7] In this system access to and use of land is controlled by a set of customary rules.

The Barabaig recognize that, to make efficient use of resources, access to grazing needs to be controlled to prevent exploitation beyond their capacity to recover. Although surface water is universally accessible to everyone, its use is controlled by rules. Routes to and from water are not to be restricted by the construction of homesteads and water sources must not be diverted or contaminated. Rights to trees are also subject to certain rules. Because of the sacred nature of some trees (*ficus spp.*) the Barabaig protect them by banning any form of use of or damage to them. Other trees that are regularly used for meetings (predominantly *acacia spp.*) are also protected in this way. Clans also have rights that cover property in land. The most important example of this is the *bung'ed*. A well becomes the property of the clan of the man who digs it. Although anyone may draw water for domestic purposes from any well, only clan members may water their stock there. Private property is recognized by the Barabaig in the form of a man's homestead and its surrounds.[8] For example, a grazing refuge (*radaneda nyega*) is marked off with thorn tree branches to protect forage near the homestead that can support small stock and sick animals that cannot keep up with the main herd. Private property also extends to trees that give shade to the household and those selected to hang beehives.

The Barabaig have a hierarchy of jural institutions that control access to and use of land, interpret customary rules and adjudicate in rare conflicts over rights and duties. Matters concerning the community as a whole are dealt with by a *getabaraku* or public assembly of all adult males. This has ultimate authority on common property rights over open rangeland. Decisions are made by consensus. If the problem has particular gravity or involves the application of sanction, a committee of the *getabaraku*, called *makchamed*, is formed. The *makchamed* deliberates in camera and hands down its decision to the assembly. Women have their own council (*girgwageda gademg*) which has involved itself with land issues. Their interest in land is related to women's special role in Barabaig spiritual life and their authority on matters involving offences by men against women. In a recent case, their council called men to account for allowing sacred land to be ploughed up for farming. Desecration of sacred land or the failure of men to protect such property is women's particular concern. At the clan level, a clan council (*hulandosht*) controls all clan property, such as *bung'eding* (plural) and wells. At the neighbourhood level, authority is vested in a neighbourhood council (*girgwageda gisjeud*). Anyone wishing to come and live in a neighbourhood can only do so with the endorsement of this council. In this way, the council effectively controls the entry of too many or too large herds and limits the potential for overgrazing.

Sanctions against those who infringe common property rules generally take the form of an order from whatever judicial moot has authority in the case. Most often the offender is simply asked to desist from the offence. If damage is done, an offender may be asked to pay a fine to the elders, or compensate an individual who has suffered as a consequence of the offence. Fines are collected by a group of youths (*orjorda*) which ensures that the transgressor complies with the ruling. Punishment for more serious offences can be greater. In the past, the women's council has been known to have issued a curse (*moshtaida*) on male offenders for breach of rules related to sacred land. This curse is most effective as it is known to bring ruin to people's lives and is much feared by all Barabaig.

Two examples highlight the rationale of the Barabaig grazing rotation and give evidence of the level of control they have over land use.

Without adherence to the movement of herds and rotation of pastures, some herders would possibly remain at the river margin (*ghutend*) where there is permanent water and persistent vegetation. However, the Barabaig understand that this would result in destruction of this resource. As herds of livestock are brought to the *ghutend* every day, whatever the season, they know that the forage there is needed by those who are watering their stock. If others are allowed to permanently graze it, this forage would soon be depleted and not be available to those who go there to draw water. This would ultimately result in destruction of the land through overgrazing and damage from concentration of hoof traffic. The Barabaig, therefore, have a customary rule that bans settlement at the *ghutend* and denies herders the right to graze the forage if they are not there to water their stock.

As water in the crater lakes of the *muhajega* is insufficient in the dry season to sustain all the livestock that congregate there, some herds have to be moved elsewhere. The wells of Lake Balangda Lelu provide alternative permanent water. However, if the pastures of the *darorajand* of Balangda are deficient it restricts access to the wells as herds cannot remain in the area for lack of grazing. Like restricted use of the *ghutend*, grazing of the *darorajeg* must be limited if pasture is to be available at the time when it is most needed. To achieve this, the Barabaig traditionally ban permanent habitation on the *darorajand*. All Barabaig accepted this. In the past they knew it was the only way to protect the grazing for the time when it would be needed in the dry season. However, with increased land pressure from encroaching farmers who occupy the *muhajega*, some Barabaig have contravened the customary rule and moved to live permanently on the Balangda *darorajand*.

This customary rule was made a by-law by the Barabaig Native Authority. Gejar, Chief of the Barabaig Native Authority circa 1936–52, recognized the threat this would have on this vital resource. He intervened by banning all permanent residence at Balangda. Today, neither customary rules nor new state laws have protected the *darorajand*. This legal vacuum has created a case of chaotic open access in which the Barabaig have had to go against their own indigenous knowledge and offend their rules and reside in numbers in the Balangda area.

Loss of land

Before colonial intervention, there was less land scarcity on the Hanang plains. Populations of people and cattle were smaller and limited by disease and marauding by the Maasai and neighbouring groups. However, with the provision of measures to improve human and animal health and the imposition of peace, the population of Hanang district has increased dramatically (Schultz 1971). Also human pressure on land in the Mbulu highlands resulted in Iraqw farmers encroaching on what was Barabaig grazing land (Schultz 1971). Although the British administration attempted to stop this it proved impossible (Schultz 1971). After a brief respite by way of the Barabaig Native Authority prior to independence, encroachment of Barabaig pasture land by small farmers from neighbouring districts of Dodoma and Singida regions also continues to the present day.

Practice of the seasonal grazing rotation means that at times in the year, and sometimes for long periods, land is free of human habitation and grazing livestock. This has led some people into thinking that some pasture land is unoccupied. It is because of this that Barabaig pastures are taken over by others and put to non-pastoral use as they are deemed to be vacant by those who want to acquire them.

The inherent fertility of the *muhajega* makes it ideal for farming. Inevitably, farmers take the best agricultural land. Just as this land has attracted Iraqw farmers, so its agricultural potential has also been noticed by developers. This has led to government appropriation of large tracts of land, including most of the *muhajega* on the Basotu plains, for an extensive wheat scheme.

TANZANIA–CANADA WHEAT PROGRAM

In response to an expected increase in demand for wheat, the Tanzanian and Canadian governments agreed to develop the Basotu plains for a state wheat scheme – Tanzania–Canada Wheat Program (TCWP). The scheme is based on a technological package reminiscent of the 'green revolution' with monocropping of hybrid varieties with large-scale mechanical equipment along the lines of prairie wheat farming in Canada. The Tanzanian partner in the venture – National Agriculture and Food Corporation (NAFCO) – set about acquiring land in the area for seven farms of 10,000 acres each, making a total combined area of 70,000 acres for the scheme (Lane and Pretty 1990).

In 1968, a Canadian mission had already identified many of the soils on the Basotu plains, and those of the *muhajega* in particular, as 'highly suitable for mechanized dryland farming' (Fenger *et al.* 1986). Yet this land was described as 'idle' (Young 1983). However, as some Barabaig herders were resident there, and compensation has also been paid to some of them for the loss of their homes, it must have been obvious the land was occupied. It can only be assumed, therefore, that those who regarded the land as 'idle' really meant that it was under-utilized and better put to a more productive purpose. This is perhaps why a decision was made by NAFCO to summarily obtain titles to a total of 100,000 acres – 30,000 acres more than they had applied for and the Barabaig had expected to accept (Lane 1991b).

The TCWP farms now cover 12 per cent of all Hanang district land (Lane and Pretty 1990). However, the significance of this loss to the Barabaig is far greater than just the land area involved. This same area represents 67.5 per cent of the land classified as 'suitable for agriculture' in the district (Tanzania government 1982). It is on this land where the greatest concentration of *muhajega* is found. If the loss of this land to the TCWP is combined with other *muhajega* lost to encroachment by farmers, the Barabaig are denied access to virtually all of the forage regime so important to pastoral production. If the spreading peri-urban development of Katesh town on the Hamit river and village expansion in the district is added to the land lost to the TCWP, together with the limited access afforded by the Mount Hanang Forest Reserve, the salt pan of Lake Balangda Lelu and tsetse fly infested bushland, it is clear that the impact of the loss is greater than would seem at first. The combined area of loss may amount to as much as 50 per cent of the area that was once available to Barabaig herds for grazing.[9] However, the loss of nearly all the *muhajega* has seriously altered the seasonal grazing rotation and had a negative effect on herd productivity and Barabaig welfare.

IMPACT OF LAND LOSS ON SEASONAL GRAZING ROTATION

The impact on the seasonal grazing rotation by the loss of *muhajega* (Figure 4.3) has forced the Barabaig to adopt a new grazing pattern, and rely more heavily on the seven remaining forage regimes.[10]

They have been forced to use *darorajand*, *heyed* and *gileud* forage when they would otherwise be rested from grazing. Further, together with the *ghutend*, it means that these are more intensely used at traditional times of use. Only the mountain (*labayd*), rift escarpment (*badod*) and bushland (*darabet*) are relatively unaffected by the loss of the *muhajega*, as these offer only sparse forage at any time, and are unable to match the productivity of the other forage regimes because of the poorer fertility and water holding capacity of soils. Changes in their grazing rotation caused by the loss of the *muhajega* have implications for the Barabaig and the Hanang plains.

ENVIRONMENTAL DESTRUCTION

The Barabaig claim that the enforced adoption of the new grazing pattern is having a negative effect on the quality of their pastures. By grazing forage regimes more intensely than before, and failing to rest them to enable them to recover, the Barabaig are becoming unwilling parties to the destruction of the land. For example, on the *darorajand*, the added grazing is causing a decline in perennial grasses and an increase in the percentage of annual herbs in the sward. The Barabaig know this herbage is less persistent in the dry season and reduces the overall productivity of the pasture.[11] A grass, called *megojiga* or 'milk grass' by the Barabaig because of its

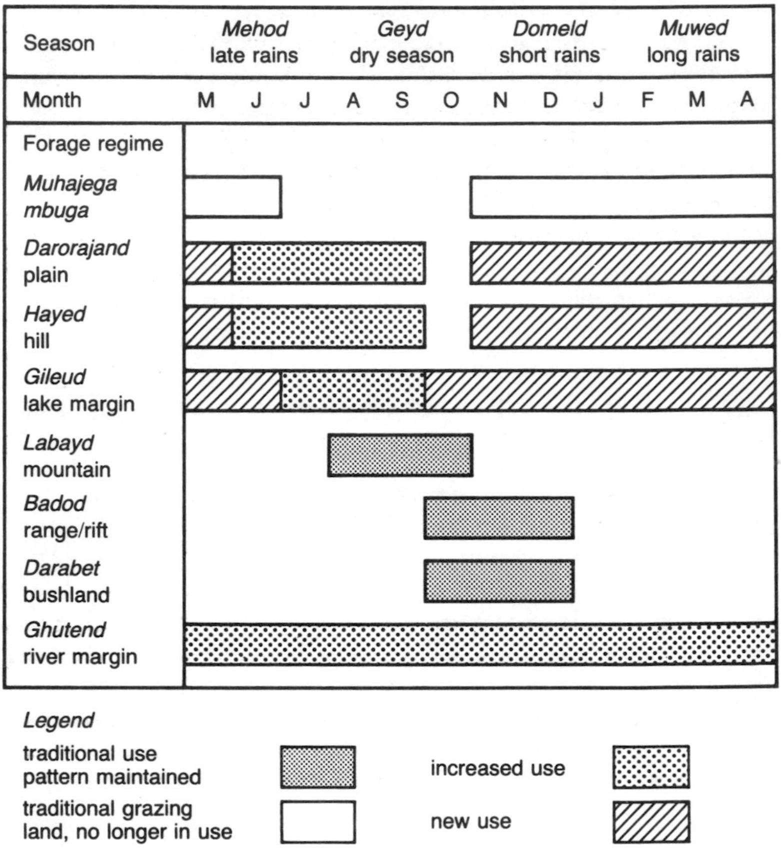

Season	Mehod late rains		Geyd dry season		Domeld short rains		Muwed long rains	
Month	M J J	A S	O N	D J	F M A

Forage regime								
Muhajega mbuga								
Darorajand plain								
Hayed hill								
Gileud lake margin								
Labayd mountain								
Badod range/rift								
Darabet bushland								
Ghutend river margin								

Legend

traditional use pattern maintained		increased use	
traditional grazing land, no longer in use		new use	

Figure 4.3 New grazing pattern
Source: Lane 1991a

milk-producing capacity, has been completely eradicated by cultivation on the farms. On the *gileud* and *ghutend*, destruction of vegetation cover by excessive hoof traffic is making land vulnerable to erosion and causing the expansion of bare soil pans. Without entering the debate about the qualitative nature of pasture changes that is well documented elsewhere (see Abel and Blaikie 1989), it is enough here to accept that the Barabaig have identified a problem in their rangeland ecology and are concerned for the future viability of their pastures under the current level of exploitation.

The Barabaig also claim that mechanized monocropping of wheat is eroding the land.[12] This is conceded by the authors of a report on the soils of the Basotu and Balangda areas commissioned by Canadian aid (Canadian International Development Agency (CIDA)). They found that the characteristics of the Basotu soils and the prevalence of high-intensity rainfall make the land susceptible to erosion by 'the removal of the natural vegetation for agricultural development [which] exposes the surface soil to accelerated erosion processes' (Fenger *et al*. 1986: 70). The report goes on to say that: 'The formation of gullies is dramatic and changes result within a few hours as large quantities of soil are moved rapidly downslope. Although gullies are the most obvious and are immediately destructive to the land base, the general downslope movement of sheet erosion may be equally detrimental' (Fenger *et al*. 1986: 72). Examples of these forms of erosion are graphically illustrated in the text. What the report fails to say is that this erosion has come about because of a failure to protect soils by the maintenance of vegetation cover that was once conserved by the Barabaig with their seasonal grazing rotation.[13]

SOCIAL INJUSTICE

The appropriation by the state of land from indigenous inhabitants inevitably raises issues of justice. This is as true for the Barabaig as it has been for other pastoral groups in East Africa (Lane and Swift 1989). Injustices have been inflicted on pastoralists through the misuse and abuse of laws to dispossess them, and in the manner of their dispossession. Further injustice comes from the suffering caused by the loss of access to those resources relied on by the Barabaig for their livelihoods.

To date, Barabaig resistance to land appropriation has failed to stop their alienation from land. Repeated complaints made to senior government and party officials over the years have not brought an effective response. Even with the creation of a separate Hanang district from what was once Barabaig division (which covers the same area as the Barabaig Native Authority), Barabaig interests have not been well enough represented by local government. NAFCO continue to expand wheat cultivation and withdrawing land from pastoral production. They have also denied the Barabaig access to grazing, water and salt by barring traditional rights of way across the farms. The cost of this denial is only exceeded by the destruction by cultivation of more than fifty of their graves on the farms, which causes the Barabaig considerable grief.

The Barabaig contend that under Tanzanian law they have

customary rights to the land of the wheat farms.[14] Being rightful occupiers of the land, they cannot be evicted without due process of law.[15] They claim NAFCO failed to adhere to that due process when they acquired the land for the farms, and thus are illegally occupying Barabaig land. A reading of the law in relation to customary rights to land suggests that there is nothing in Tanzanian law that denies pastoralists' communal claim to rights over common grazing land (James and Fimbo 1973; James 1971), and their rights are in no way inferior to individual claims to private land that have succeeded in the courts.[16] In theory then, the Barabaig should be able to successfully assert their rights to possession of common land, and defend it from wrongful alienation.

In 1984, some residents of Mulbadaw village took NAFCO to court and contested the appropriation of land for the Mulbadaw wheat farm.[17] After the initial High Court decision in their favour, the Appeal Court ruled against the villagers on a technicality.[18] However, the more fundamental issue of whether the Barabaig have customary title to the land was not resolved. With continued expansion of cultivation, the Barabaig have returned to the courts. In a current case, they are disputing NAFCO's occupation of the 30,000 acres acquired in excess of the 70,000 acres originally requested. They want full compensation for the loss of this land. Progress in this case was set back by the issue of a government order extinguishing any customary claim to land in the vicinity of the TCWP farms.[19] This order was made retroactive to a date prior to the lodgement of the Barabaig application to the courts. Barabaig lawyers argue that the order is unconstitutional as it attempts to extinguish entrenched aboriginal rights of all Tanzanians without following proper legal procedures. The human rights monitoring group, Africa Watch, believes this order breaches several provisions of the African Charter on Human and People's Rights, to which Tanzania is a signatory. They also believe its retroactive effect contravenes a basic principle of international human rights law and natural justice (Africa Watch 1990).

In socio-economic terms, the Barabaig can show that the reduction in the resource base by the withdrawal of the *muhajega* is being reflected in a decline in cattle numbers and a worsening in the quality of their lives. Following interviews with over five hundred pastoralists in the Gehandu and Mogitu villages areas adjacent to the wheat farms, cattle herds in 1988 have declined to a third of their size in 1981 as a result of mortalities attributed to the stress of reduced grazing. Also Hanang district health statistics show that child health and nutrition are significantly worse in communities near the farms than found in other areas beyond them (Gitagnod 1988). As yet, it is impossible to reveal a direct link between the TCWP and lack of Barabaig well-being. However, indications are that, whatever the scheme is achieving in production of wheat, it is failing to advance the welfare of neighbouring Barabaig communities.

UNSOUND ECONOMICS

It is argued by supporters of the TCWP that the scheme has a national importance that exceeds the problems for the Barabaig. However, this apparent justification becomes less tenable under examination of the scheme's economic performance. Although the farms satisfy nearly half the national demand for wheat, they do not satisfy the major goal of helping Tanzania become food self-sufficient. Wheat only makes up 2–3 per cent of food crop consumption (Carter *et al.* 1989). Moreover, most of the demand comes from wealthy urban dwellers and not the rural poor (Carter *et al.* 1989). Because of the high-technology and capital-intensive nature of production on the farms they have created more of a dependency on Canadian aid than self-sufficiency in food production (Freeman 1984).

The economic performance of the TCWP has also been challenged by both internal evaluations and independent assessments. CIDA and the Tanzanian government continue to rely on past favourable financial assessments that showed the scheme to have a positive cost–benefit ratio and a 40 per cent internal rate of return (Stone 1982). However, in 1982 the economic realities of the TCWP came to light. Two internal evaluations revealed that, while the TCWP appeared to be profitable in financial terms, it was uneconomic in that the costs exceeded the benefits (Prairie Horizons 1986; Michael Mascall and Associates 1986). In 1986, an independent study was done with the purpose of testing the efficiency of resource use and to find out whether it makes economic sense to grow wheat by the methods employed on the scheme compared with smallholders using oxen, and relative to the cost of direct importation of wheat from the world market (Carter *et al.* 1989). The study concluded that the scheme had a negative net financial profitability and a negative benefit–cost ratio. For the first time, the 'sunk cost' of

Canadian aid was taken into account and given an opportunity cost to the Tanzanian economy. When 'all costs and benefits from the point of view of society as a whole' and not just those 'costs and returns as faced by the individual or firm' were taken into account it was found that the scheme was unprofitable (Carter *et al*. 1989: 17). This led to the conclusion that the TCWP 'is not economically viable nor does it make effective use of domestic resources in saving foreign exchange for Tanzania to use large mechanised wheat production to satisfy the domestic demand for wheat' (Carter *et al*. 1989: 24).

PERSISTING WITH OLD ORTHODOXY

The plight of the Barabaig is typical of a wider problem for pastoralists throughout Africa. Common land tenure systems have long been thought incapable of efficient land use (Lane and Swift 1989). Evidence of rangeland erosion since colonial times has been blamed on the irrational behaviour of pastoralists who accumulate cattle in excess of their economic needs for reasons of social prestige. Credence was given to this view by reference to the 'cattle complex' that dwelt on the cultural importance of cattle in traditional pastoral societies (Herskovits 1926). This anthropological legacy has since been reinforced by the 'tragedy of the commons' scenario which posits that individual herders have no incentive to restrict stock numbers, and the herding of private animals on communal pastures will inevitably lead to overgrazing and land degradation (Hardin 1968).

Despite the time that has elapsed since Hardin first published his essay more than twenty years ago, and a mounting challenge to it on both theoretical and empirical grounds (see Runge 1986), it has remained a powerful force in the minds of government officials and aid agency personnel to this day. This is revealed in the continued espousing of the 'old orthodoxy' that pastoralists have too many cattle and that, because of uncontrolled use of the commons, they will inevitably destroy the land through overgrazing (Lane and Swift 1989: 1). Ignorance of the nature of pastoral production systems and the value of traditional pastoral common land tenure arrangements, together with acceptance of flawed theories have led to policies that work to undermine traditional pastoralism and support for privatization of common land.

Many studies have shown that pastoralists do not have excessive numbers of livestock. Even so-called 'pure pastoralists' who practise a minimum of cultivation rarely have enough livestock on which to subsist on a pastoral diet (Swift *et al.* 1989). The average total number of cattle in a Barabaig household at Balangda is only 60 animals, or 6 per person. As only 30 per cent of these are milch cows, it means that each person receives milk from only two cows. As Barabaig cows provide only just over 1 litre of milk a day for human consumption and their lactation is only for around eight months, current herd size is clearly not enough for everyone to be self-sufficient in milk throughout the year. That is why the Barabaig and other Tanzanian pastoralists cultivate crops to supplement their diet.

Despite the reality of pastoral existence in Tanzania, new thinking on pastoral development is not accepted by planners. The Comment column of the Tanzanian *Daily News*, an editorial column which frequently expresses official government policy, reiterated the claim, made in the past by the British colonists, that pastoralists keep excessive numbers of cattle which leads to over-grazing: 'Therefore the Party [CCM] and government leaders at all levels must educate and persuade cattle keepers in particular to de-stock in order to improve their own lives and preserve the soils and pastures that sustain their animals.'[20] In accepting this reasoning, the government is ignoring Barabaig needs for survival and the fact that land degradation comes more from the land alienation than excessive numbers of stock.

An almost verbatim expression of the 'tragedy of the commons' scenario is found in an article detailing measures for 'proper management of rangelands' by an official of the Tanzanian Ministry of Agriculture and Livestock Development:

> this practice of grazing private livestock on communal land constitutes the single major constraint to improved manage-ment of the natural pasture lands. The inevitable result of this system of livestock production is that the cattle owners keep excessive numbers of livestock which in turn leads to over-grazing, soil degradation, low fertility and high mortality rates. However, in order to allow for the best possible care of the agricultural land in the future, users will be allocated land on the basis of lease-hold, thus ensuring that they get full legal protection ... restriction of animal numbers to any reasonable balance with the forage resource has proved difficult due to lack of land ownership rights and communal land ownership.
>
> (Bilali 1989)[21]

Despite Tanzania's ranking as one of Africa's highest cattle resource countries, livestock production per capita has declined over the years (Mbilinyi *et al.* 1974). In response to this, in 1982 the government published a National Livestock Development Policy that remains a guide to the direction of development in the livestock sector to the present day (Tanzania government 1982). The preamble acknowledges that 99 per cent of livestock were in the hands of traditional producers, and that there is enormous potential for increased production. Yet the measures given to convert this wealth exclude traditional pastoralism from support because it is regarded as backward and unproductive:

> The long-term objective is therefore to bring about changes in traditional producers' attitude and practices thereby increasing productivity to the level where this sector evolves into the modern sub-sector.
>
> (Tanzania government 1982: 4)

The 'modern sub-sector' envisaged in this context is made up of private and state ranches that do not include much pastoral involvement (Mustafa 1986). It seems that a necessary aspect of modernization is the transformation of the traditional sector, which invariably means the destruction of pastoralists' traditional land use management systems based on common land tenure.

The policy also suggests encouraging pastoralists to be moved out of overstocked areas to 'lower-stocked or virgin areas suitable for livestock' (Tanzania government, 1982: 4). However, it does not consider the alternative of resettlement of farmers out of pastoral areas so that pastoralists can benefit from the full production potential of the land. The policy concedes that 'present users need to be given assured rights over land they are using' (Tanzania government 1982: 5). Although it recognizes the existence of communal land tenure, it implicitly promotes private property as a means of giving land holders security so that they will 'feel confident that their investments of effort and money will be beneficial to them and their families' (Tanzania government 1982: 5). It also leaves unexplained how pastoralists might attain security of land when rights of access and use are conferred on the whole community. If they are to rely on existing authorities to administer these rights and protect common land, then the past performance of government would suggest that the prospects for pastoral land rights are not encouraging.

A POSSIBLE WAY FORWARD

As a counter to the old orthodoxy and as a means of averting what is happening to the Barabaig and advancing the prospect of sustainable development on the Hanang plains, changes are needed in the way development is being conducted in pastoral areas of Tanzania. This is not so much to single out pastoralists for preferential treatment, but simply to address the failures that have penalized them in the past. An important first step in this process is to acknowledge the value of pastoral common land tenure systems as a means of securing pastoral livelihoods and of conserving natural resources.

Consistent with the nation's socialist ideals and expressed commitment to social equity, the needs of the Barabaig should be given high priority. These needs have been unequivocally expressed in an open letter to the Canadian people from the Barabaig (in Paavo 1989). In this letter they are asking that their customary rights to the land they occupy be recognized and access to that land secured. For this to happen, it is vital the court case is allowed to be judged on its legal merit. This excludes interference from the executive (Africa Watch 1990). Without a legal precedent confirming pastoralists' customary communal rights to land, the future of the Barabaig and the land on which they depend remain threatened. This has implications beyond the Barabaig in Hanang district. It is a matter of utmost concern to other pastoral groups, and other communities, who have common land tenure systems. If the Barabaig case is not successful, then legislation needs to drafted so as to allow indigenous inhabitants to better defend their rights and more effectively contest wrongful alienation from their land.

Once land security is attained, a new direction for development needs to be found that more effectively supports pastoralism. This will require revision of policies that not only express the potential of livestock production but also caters specifically for pastoralists' needs. The condemnation of common land tenure should be replaced by measures that acknowledge the value and importance of these systems. Development projects should be designed to make better use of pastoralists' indigenous knowledge and skills in resource use. One way of doing this is to build up pastoral communities' capacity to manage local assets and resources. More infrastructure is needed to improve stock health and livestock and crop productivity to levels that will better enable the Barabaig to exploit the market for the benefit of both themselves and the wider national economy.

At the district level, new initiatives should start with local structures. Participatory research methods should be employed to uncover constraints on livestock production and welfare needs of pastoralists. The same participatory processes should be used to empower local communities to overcome these constraints. This will require government officials and aid agencies taking more interest in and working more closely with local communities. To facilitate this, it will be necessary to identify and support a unit of social organization that has its roots in indigenous culture. Through this institution, the Barabaig can be more fully consulted and their views taken into account in developments that affect them. 'Area-based' social services should be provided to more effectively reach semi-nomadic pastoral communities (Swift 1989).

If these measures are taken, much of the human suffering and damage to the environment can be repaired. The Barabaig still offer an opportunity to support a coherent traditional and sustainable land use system that could show the way for development in other pastoral areas of Tanzania and replace the old orthodoxy that has to date blighted pastoral development throughout much of Africa.

NOTES

1 Many non-Barabaig Tatoga are called 'Barabaig' by people (including government officials) who do not know the history and details of Tatoga society. Another name 'Mangati' is also widely used for Tatoga people. It is a disparaging term that comes from the Maa words il Mang'ati, meaning enemy (plural).
2 It is impossible to be more accurate about their number as ethnic origin has not been recorded in a Tanzanian census since 1956.
3 Sometimes called 'Mangati plains'.
4 Balangda is the Tatoga word for salt.
5 This model was developed with the help of the Barabaig elders of Balangda division. It is a summation of their individual strategies. It is to be remembered that it represents an historical view as movement in the district has become severely restricted in the last twenty years.
6 There are also sometimes compelling social reasons for Barabaig movements and location of habitation, but these are not the subject of this discussion.
7 A household can be made up of a number of homesteads that can be in a group or spread over a great distance.
8 Sometimes, but rarely, a household is headed by a woman. Even where a

household is headed by a man, rights of 'ownership' to animals of the household herd are vested in members of the family, including women.

9 This can only be quantified by an aerial land use survey of the whole district.

10 There are a few very much smaller *muhajega* on the Basotu and Barabaig plains, but they are not large enough to sustain many animals and their contribution to cattle production is insignificant compared with those lost to the wheat scheme.

11 Whilst this has yet to be quantified there is no reason to doubt Barabaig assessment, as they have, like other traditional pastoralists, an intimate and detailed knowledge of rangeland ecology.

12 An Open Letter to the Canadian People for the Barabaig of Tanzania, in Paavo (1989).

13 This is not to deny that convergence of livestock trails at water points has caused some minor soil erosion in these localities.

14 Customary rights to possession were conferred on indigenous occupiers of land by the 1923 Land Ordinance which remains law to this day.

15 Conditions for the appropriation of land are set out in the 1967 Land Acquisition Act.

16 *Lagwen Irafay and 19 Others v. Nangwa Village Council*, Civil Case No. 4 of 1982, High Court at Arusha.

17 *Mulbadaw Village Council & 67 Others v. NAFCO*, Civil Case No. 10 of 1981, High Court of Tanzania, Arusha.

18 *NAFCO v. Mulbadaw Village Council & Others*, Civil Appeal No. 3, Court of Appeal, 1985.

19 Extinction of Customary Land Rights (Amendment) Order, 1989, Government Notice No. 260, 28.7.89.

20 *Daily News*, 21 January 1982.

21 Despite a disclaimer by the editor of the above publication that the views expressed by contributors of articles are not necessarily those of the governments of SADCC member states, it is clear from other official sources that it is an accurate reflection of official thinking.

REFERENCES

Abel, N. and Blaikie, P. (1989) 'Land degradation, stocking rates and conservation policies in the communal rangelands of Botswana and Zimbabwe', *Land Degradation and Rehabilitation*, 1: 101–23.

Africa Watch (1990) 'Tanzania: executive order denies land rights – Barabaig suffer beatings, arson and criminal charges', *News from Africa Watch*, London.

Baxter, P. (1989) *Property, Poverty and People: changing rights in property and problems of pastoral development*, Department of Social Anthropology and International Development Centre, University of Manchester.

Bilali, A. (1989) 'Management of pastures and grazing lands in Tanzania', *Splash* 5 (2): 6–7, Maseru: SADCC.

Borgerhoff-Mulder, M., Sieff, D. and Merus, M. (1989) 'Disturbed ancestors: Tatoga history in the Ngorongoro Crater', *Swala*, 2 (2): 32–5.

Carter, C., Frank, D. and Loyns, R. (1989) 'Wheat in African development: the case of Tanzania', paper presented to Annual Conference of the Canadian Association of African Studies, Carlton University, Ottawa.

Conway, G. and Barbier, E. (1988) *After the Green Revolution*, London: Earthscan.

Fenger, M., Hignett, B. and Green, A. (1986) 'Soils of the Basotu and Balangda Lelu areas of northern Tanzania', Ottawa: Agriculture Canada and CIDA.

Freeman, L. (1984) 'CIDA and agriculture in East and Central Africa' in J. Barker (ed.) *The Politics of Agriculture in Tropical Africa*, Beverly Hills: Sage.

Gitagnod, D. (1988) 'What value a cow?' paper given at workshop 'Pastoral land tenure in East Africa', Arusha, Tanzania, December 1988, IDS, University of Sussex.

Hardin, G. (1968) 'The tragedy of the commons', in G. Hardin and J. Baden (eds) *Managing the Commons*, San Francisco: W.H. Freeman.

Harrison, P. (1987) *The Greening of Africa: breaking through in the battle for land and food*, London: Earthscan.

Herskovits, M. (1926) 'The cattle complex in East Africa', *American Anthropologist*, 28: 230–72, 361–80, 494–528, 630–64.

James, R. (1971) *Land Tenure and Policy in Tanzania*, Nairobi: East Africa Literature Bureau.

James, R. and Fimbo, G. (1973) *Customary Land Law of Tanzania*, Nairobi: East Africa Literature Bureau.

Lane, C. (1991a) 'Alienation of Barabaig pasture land: policy implications for pastoral development in Tanzania', D. Phil thesis, Sussex: Institute of Development Studies, University of Sussex.

Lane, C. (1991b) 'Wheat at what cost? The Tanzania–Canada Wheat Program', in J. Swift and B. Tomlinson (eds), *Conflicts of Interest: Canada and the Third World*, Toronto: Between The Lines.

Lane, C. and Pretty, J. (1990) 'Displaced pastoralists and transferred wheat technology in Tanzania', *Gatekeeper*, SA20, London: IIED.

Lane, C. and Swift, J. (1989) 'East Africa pastoralism: common land common problem', *Issues Paper* 8, IIED Drylands Programme, London.

Makec, J. (1988) *The Customary Law of the Dinka People of Sudan*, London: Afroworld.

Mbilinyi, R., Mabele, R. and Kyomo, M. (1974) 'Economic struggle of TANU government', in G. Ruhumbika (ed.) *Towards Ujamaa: 20 years of TANU leadership*, Nairobi: East Africa Literature Bureau.

Michael Mascall and Associates (1986) 'Report on the evaluation of the benefit/cost report and production cost analysis of the Tanzania wheat project reports by Prairie Horizons Limited'.

Mustafa, K. (1986) 'Participatory research and the "Pastoralist Question" in Tanzania: a critique of the Jipemoyo project experience in Bagamoyo district', unpublished Ph.D. thesis, University of Dar es Salaam.

National Research Council (1986) *Proceedings of the Conference on Common Property Resource Management*, Washington, DC: National Academy Press.

Netting, R. (1978) 'Of men and meadows: strategies of alpine land use', *Anthropology Quarterly*, 45: 132–44.

Paavo, A. (1989) 'Land to the stealer', *Briarpatch*, 18 (7): 22–4.

Pearce, D., Markandya, A. and Barbier, E. (1989) *Blueprint for a Green Economy*, London: Earthscan Publications.

Prairie Horizons Ltd, (1986) 'Final report of the benefit/cost team on the Tanzania wheat project submitted to the Natural Resources Branch, CIDA'.

Raintree, J. (ed.) (1987) *Land, Trees and Tenure*, Nairobi: ICRAF.

Runge, C.F. (1986) 'Common property and collective action in economic development', *World Development*, 14 (5): 623–35.

Saitoti, ole T. (1986) *The Worlds of a Maasai Warrior: an autobiography*, New York: Random House.

Schultz, J. (1971) *Agrarlandschaftliche Veranderungen in Tanzania*, Munchen: Welt-Forum.

Stone, J. (1982) 'Project evaluation: a case study of the Canada–Tanzania wheat project', Guelf: University of Guelf.

Swift, J. (1989) 'Getting services to nomadic pastoralists in Kenya', Nairobi: UNICEF.

Swift, J., Toulmin, C. and Chatting, S. (1989) 'Providing services for nomadic peoples: a review of the literature and annotated bibliography', WHO/UNICEF.

Tanzania government (1982) *The Livestock Policy of Tanzania*, Dar es Salaam: Ministry of Livestock, Government Printer.

Wilson, Mcl. G. (1952) 'The Tatoga of Tanganyika', *Tanganyika Notes and Records*, no. 33.

Young, R. (1983) *Canadian Development Assistance to Tanzania*, Ottawa: North–South Institute, p. 67.

5

THE *ZANJERAS* AND THE ILOCOS NORTE IRRIGATION PROJECT: LESSONS OF ENVIRONMENTAL SUSTAINABILITY FROM PHILIPPINE TRADITIONAL RESOURCE MANAGEMENT SYSTEMS

Ruth Ammerman Yabes[1]

Water is a critical production input in sustainable agricultural development. A major strategy for Asian agricultural growth has been the creation or improvement of irrigation systems through water development and/or management activities (Tamaki 1977). Most of these efforts may be characterized as falling into one of three categories: (1) agency-based, bureaucratic systems, (2) community-based, communal systems, or (3) joint bureaucratic-cum-communal systems (Coward 1980). Bureaucratic systems may be defined as large-scale, and centrally planned and/or controlled; and communal ones as small-scale, and independently or locally planned and controlled (Barker *et al.* 1984).

Terms such as traditional, communal, community, indigenous, local and farmer-managed are used in irrigation literature to describe irrigation systems that are operated with *little or no input from the government or outside agencies*. Each of these terms has different connotations depending on the perspective taken. For the purposes of this chapter, the term 'communal' will be used to describe irrigation systems where the system's water users are the same ones who make decisions about operation, maintenance and conflict management, since this is the terminology most commonly employed in Philippine irrigation literature (Yoder 1986; Eggink and Ubels 1984).

In South-east Asia the *subaks*, or irrigation societies, of Bali, Indonesia, the *muang-fai* irrigation systems of northern Thailand and the *zanjeras* of the Philippines are communal irrigation systems with traditional water resource management systems. *Zanjeras* are a well-known example of communally created and operated irrigation in the north-western part of the Philippines. One of their central purposes is to 'procure a stable, reliable supply of water, which can increase crop production in some cases by more than half' (Lewis 1971: 128). An equally important goal is to increase access to the use of arable land, a very scarce resource in Ilocos Norte, the home province of the *zanjeras*.

Zanjeras are groups of users with organizational structures and leadership rules which allow the users to be in charge of their irrigation environment, and to manage that environment with a considerable degree of effectiveness within the natural limits of the habitat and the technical limits of the physical apparatus at their disposal. Most *zanjera* irrigation activity is undertaken without government assistance or intervention.

However, despite the strength of *zanjeras* throughout Ilocos Norte, in 1978 the National Irrigation Administration (NIA) began the design of a 'modern' large-scale irrigation scheme, the two-phase Ilocos Norte Irrigation Project (INIP). This project was designed to bring in additional water supplies, and to introduce hydropower to the region. The *zanjeras* occupied a part of the setting for INIP, but they were unfortunately largely ignored by the original project planners. It might be said that the *zanjeras* were not so much unknown as undervalued – known but disregarded by NIA and INIP's Japanese planners and donors. Yet *zanjeras* have a long history which cannot be ignored. Nor did the *zanjeras* ignore INIP: *zanjera* officers and members protested loudly against the intervention by NIA and its irrigation project, which was developed with almost no involvement of the *zanjeras*.

This chapter examines the origins and organization of the *zanjeras*, and includes a discussion of the organizing principle of the *atar*, or membership share. A review of *zanjera* rules reveals who participates in these irrigation societies, and in what ways. The benefits of *zanjera* participatory activities, which include productivity, equity and environmental sustainability are examined. Challenges to *zanjera* activities are also explored.

The second part of the chapter discusses the intervention by NIA and the INIP project in an area already filled with existing *zanjera* irrigation systems. Project plans, and the two different planning approaches undertaken to INIP – the standard one (engineering-based) and a more participatory one – are described. Reasons for *zanjera* resistance to the project are examined. The next section considers the environmental problems of the INIP project, including flooding, soil erosion and siltation, and the environmental consequences for the *zanjeras*. The last section identifies what needs to be emphasized or avoided when implementing participatory planning approaches to promote environmental and agricultural sustainability in traditional resource management systems.

ZANJERA ORIGINS

The term *zanjera* is derived from the Spanish word *zanja*, which means ditch or conduit. *Zanjeras* are organizations that build and maintain irrigation ditches. They are known in the Philippines and abroad for their enduring capacity to manage gravity-fed communal irrigation systems, and for their rules and regulations governing water allocation and distribution, system operation and maintenance, and conflict management.[2] For example, when one *zanjera* in the project area was organized in the 1850s there were conflicts about water allocation and distribution, so proportional weir structures were built by the *zanjera* to mediate these conflicts. Other *zanjeras* had been operating for up to two centuries prior to the arrival of the INIP regional project. The systems vary from simple to complex, depending on their size and physical condition, and whether they are located on rough terrain or on plains. Some of the larger *zanjeras* are divided into smaller units called *gunglos*.

There are between 1,000 and 1,200 *zanjeras* in Ilocos Norte (Figure 5.1).[3] They range in size from one to a thousand hectares. Scattered throughout the province, their total land area in Ilocos Norte has been estimated from a low of 15,000 hectares by Christie in 1914 to 32,592 hectares by Thomas in 1978.[4] As of 1984 the Philippine Ministry of Agrarian Reform identified 10,664 'farmer-tillers' cultivating an aggregate area of 7,686.33 hectares in the INIP project area (Ministry of Agrarian Reform 1984) which contains approximately 180 to 200 *zanjeras*, according to INIP project managers.

Ilocos Norte farmers operate in a demanding physical environment. Over two-thirds of the province is mountainous, with only 12.4 per

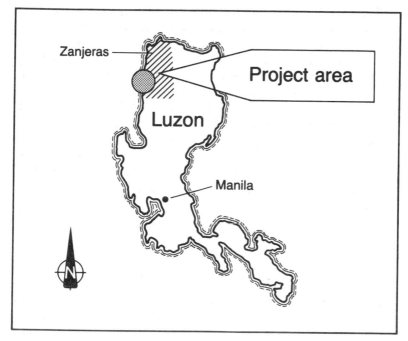

Figure 5.1 Location of *zanjeras*
Source: Yabes 1990

cent of the total land area farmed and available for cultivation, most of it located on coastal plains. The Ilocos region is bordered to the west by the South China Sea and by the Cordillera Central mountain range to the east, with elevations ranging from 5,000 to 10,000 feet. Ilocos Norte experiences a marked rainy season from mid May through mid October, and a dry season which is divided into a cool period from mid October through February and a hot period from March until mid May. The onset of the rainy season varies throughout the province. Towards the eastern mountains in central Ilocos Norte the rains may begin as early as mid April (Thomas 1978; Lewis 1971).

Keesing found reports of the existence of irrigated agriculture in Ilocos Norte in the writings of Spanish priests from as early as 1630.[5] There is neither agreement on nor any record of whether these irrigation activities pre-dated the Spanish, or were introduced and developed by them (Siy 1982).

Little detail remains concerning irrigation techniques practised in this early period (Keesing 1962). Christie provides some information, in his 1914 article, about the situation at the time. Calling the *zanjeras* 'irrigation societies', he gives a glimpse of the crude physical nature of the *zanjera* system:

> Neither town has dams of a permanent nature. Diversions are made usually by temporary dams of bamboo and rock from 0.5 meter to 3 meters high. These are crudely constructed, and are either completely destroyed each year or require considerable repairing. Some of the 27 ditches in these towns have no head gate nor wasteway provisions, and as a consequence their channels have been cut so deep as to leave portions of the lands they once watered above water.

(p. 99)

ZANJERA ORGANIZATION

Zanjera diversion dams, made of brush, sticks, stones and leaves, wash out 5 to 10 times each year, depending on the weather. Water from these dams flows into earthen canals which must be regularly cleaned of grass, debris and silt. Work parties are formed by *zanjera* members, on either a regular or an emergency basis, depending upon circumstances, to construct and repair dams and clean canals. Individual work assignments vary according to the difficulty of the tasks undertaken, from as little as one to two hours of work by a few persons for canal cleaning, to up to 6 to 9 days of continuous work by over 400 men for completely re-building the largest of the dams. Members may contribute from 5 to 80 days of labour per year, depending upon the severity of the weather and other factors.[6]

One type of *zanjera* is built on cultivated lands, by people who do not own the land, in exchange for usufruct rights on part of the newly created *zanjera*.[7] The newly irrigated land is divided between landowners and the *zanjeras*, with the latter obtaining the use of between one-third and five-sixths of the total irrigated area (Siy 1982). The arrangement is formalized in *biang ti daga* agreements (Iloko for 'sharing of riceland').[8] *Zanjeras* are obligated to provide the landowners with water in perpetuity, and risk forfeiting their rights to their portion of the land should they fail.

THE *ATAR*, OR MEMBERSHIP SHARE

Zanjeras with *biang ti daga* arrangements work on the basis of the *atar* or membership share.[9] The *atar* is central to understanding

110

how this type of *zanjera* functions and is one of the key principles behind member participation in *zanjera* activities.[10]

The word *atar* means 'share'. It is at one and the same time a unit measure of land, as well as a system of rights and responsibilities.[11] The number of *atars* in a *zanjera* is fixed when it is initially organized. The *atars* do not change subsequently unless the *zanjera* physically expands its area. *Atars* are neither uniform nor specific, and so they vary in the land area they represent. For example, an *atar* might be one-quarter of a hectare in one *zanjera* and two hectares in another.

Zanjeras with *atar* memberships base their system of rights and obligations to land, labour and water on this division. Labour and material resources required to be contributed are proportional to the number of *atars* held. *Zanjera* Danum,[12] for example, requires every *gunglo* to provide one member for each *atar* for *dagup* labour towards repair and maintenance.[13] The landowners associated with or who are owner-operators in *Zanjera* Danum are also required to make contributions of money and food based on the *atar*.

ORGANIZATIONAL STRUCTURE OF THE *ZANJERAS*

The organizational structure of *zanjeras* reflects their physical complexity and size. Smaller, simpler *zanjeras* have only a leader (*panglakayen*) and a secretary/treasurer. Larger ones often have several types of officials, such as president, vice president, secretary, treasurer, auditor and a board of directors. Sometimes there are appointed advisers and cooks.[14] In the research area, several *zanjeras* had officers who also held concurrent political positions in local government, such as municipal council member, or vice mayor.

Zanjera Danum is an example of one of the more complex *zanjeras*. It has two sets of functionaries. One set of officers governs everyday water management activity. These individuals are elected from among the membership, and therefore may be tenants, lessees or owner-operators. This set of officers has three levels which correspond to the physical arrangement of the irrigation system. The lowest level encompasses each of Danum's thirty-three *gunglos* (sub-sections) which are coordinated by *panglakayen*, with the help of *gunglo* secretaries and sometimes an additional vice-*panglakayen* or other *gunglo* officer. At the next level, the *zanjera* is physically divided into three sections which correspond to three lateral canals. A *segundo cabecilla* heads each of these sections. At the highest

111

level, the overarching leadership roles consist of the *mayor cabecilla*, secretary, sub-secretary and treasurer.

A second and separate set of officers represents the landowners of Danum. This group includes a president, vice-president, secretary and board of directors. The president's position is traditionally held by a descendant of one of the four original landowning families in the *zanjera*.[15]

ZANJERA PARTICIPANTS IN IRRIGATION ACTIVITIES

People participate in *zanjeras* in different ways according to their membership status and leadership position.[16] Lewis identifies two types of members and two types of non-members in *zanjeras*. The two types of *zanjera* members are *atar* shareholders and those with *inkalian* lands, or the 'canal builders'. The two types of non-members include the landowners in the *biang ti daga* arrangement, and the *inkapulo* group, or the 'water buyers' who pay a 10 per cent share of their harvest to buy water from a *zanjera*.[17] Differences among the *zanjeras* and their members within INIP project boundaries may be examined along three lines – location, landholding and gender.

Location

One of the more common differentiating characteristics of water users is their position in the upstream/downstream dichotomy in an irrigation system. In the *zanjeras* with an ideal *atar* system, this does not arise. Each of the original water users has an *atar* parcel in each of the zones (upstream, midstream and downstream) located in the *zanjera*.[18] Thus in *zanjeras* with an intact *atar*-based system the upstream-downstream differential does not exist within a particular *gunglo*. This evenness in holdings encourages participation in *zanjera* activities because cause and effect are more broadly distributed. For example, a *zanjera* member stealing water for his upstream parcel denies himself downstream.

In practice, however, the *atar* share has been divided into fractions in a growing number of *zanjeras* due to inheritance and land sales. Consequently some *atar* parcels have been divided between or among different persons. The result is that not all *atar* shareholders have both upstream and downstream parcels.

Landholding

Access to land is very important to *zanjera* members. However, land ownership and tenure patterns in Ilocos Norte are hard to categorize. In 1903 the average farm size in Ilocos Norte was 0.62 hectares.[19] In 1982 the average was 1.6 hectares in the province as a whole, and was estimated to be 0.824 hectares in the INIP pilot project area (Visaya 1982). These small landholdings are fragmented into several even smaller parcels, because of the *atar* zone system and a pattern of ownership where land is bought and sold to pay for debts and children's education. A 1984 Ministry of Agrarian Reform (MAR) survey of the tenure status of farmers in the pilot and Phase I areas of INIP indicates that of the 10,664 farmers in the INIP area covering 7,686 hectares, 30 per cent are share tenants, 40 per cent are lease holders, 9 per cent are amortizing owners and 21 per cent are owner-operators (Ministry of Agrarian Reform 1984). Transient farmers comprise 35 per cent of the farming households in the pilot area (Visaya 1982).

These figures provide a glimpse of the general tenure status of farmers in the project area, but do not accurately reflect how many of these farmers are simultaneously owner-operators, lessors, lessees and tenants of these parcels. *Zanjera* members highly value their membership as a means of access to land and water rights, given the small, average farm size in Ilocos Norte, and the fact that many of these farms are divided into several even smaller parcels.

Gender

The majority of members in most of the *zanjeras* surveyed are men. There are female members in only 6 of the 45 *zanjeras*, and these are usually widows of deceased male members. Women did not participate in water management functions, or hold office. None of the *zanjeras* with female members allowed women to contribute labour during work parties. Instead women were expected to be non-working members who pay an annual fee in lieu of contributing labour.

ZANJERA PARTICIPATION IN IRRIGATION ACTIVITIES

Zanjeras are vital organizations. *Zanjera* members participate in all aspects of the communal irrigation system. *Zanjera* resource management activities and their rules reflect the physical and organizational characteristics and requirements of each system. These activities and rules contribute to the long-term environmental sustainability of the *zanjeras*. The extent and nature of each member's participation depends on the type and range of tasks undertaken by each *zanjera*. The area to be irrigated, the complexity of the physical environment (topography, water source, soils), and the stability of the irrigation structures (temporary versus permanent) are the governing factors.[20]

Zanjera members participate in most of the four kinds of irrigation management activities suggested by Uphoff (1986), dealing with: (1) water use; (2) control structures; (3) internal organization; and (4) external organization.[21] Emphasized here will be those *zanjera* activities which have implications for resource management and environmental sustainability.[22]

Participation in water use activities

Uphoff defines water use activities to include: (1) acquisition; (2) allocation; (3) distribution; and (4) drainage (Uphoff 1986). Discussions with *zanjera* officers revealed extensive farmer participation in the first three categories with little or no attention to the fourth (not discussed here).

Acquisition

Zanjeras acquire water through their temporary brush dams. Construction and repair of these dams are central to the operation of the *zanjera* irrigation systems. Thus, most *zanjeras* require all members to help build brush dams or to pay instead a substantial annual fee to be a non-working member. For example, in Danum a non-working member must pay an annual fee of P140 (US$ 7), and in some of Danum's *gunglos*, an additional P150 (US$ 7.50) annual fee to the *gunglo*.[23] Members contribute labour, materials and sometimes cash as needed for building these brush dams in order to divert water for irrigation.

Allocation and distribution

In order to ensure sustained delivery and equitable distribution of irrigation water, *zanjeras* have water allocation rules for water-sharing arrangements within their own groups as well as with other *zanjeras*. These include dates and times for water delivery, labour schedules and assignments for rotation activities, and various punishments and fines for water stealing and other distribution-related offences.

Water distribution activities vary between the wet and dry seasons in response to variations in the water supply volume. Forty of the 45 *zanjeras* surveyed in the INIP project area indicated that they use continuous flow irrigation in the wet season when water supplies are relatively abundant.[24] Farmer participation during continuous irrigation periods focuses more on responding to excess water emergencies rather than on setting up distribution schedules, as in the dry season. However, in the dry season, with scarce water supplies, over three-quarters of the *zanjeras* (37 out of 45) said they use rotational (*squadra* or *cuadra*) irrigation, while about one-ninth (5 out of 45) use continuous irrigation since they are supplied by perennial springs or drainage.

PARTICIPATION IN CONTROL STRUCTURE ACTIVITIES

Zanjera members are actively involved in the design, construction, operation and maintenance of their canals and structures. This participation contributes to the long-term sustainability of their systems. Field interviews with *zanjera* officers and individual farmers identified a range of types of participation in control structure irrigation activities. Uphoff (1986) lists the following participatory activities for control structures: (1) design; (2) construction; (3) operation; and (4) maintenance.

Design and construction

Zanjera brush dams vary in terms of size and type of construction materials in response to variations in river volume and velocity and availability of materials. One brush dam design example is from *Zanjera* Pam-pan-niki and *Zanjera* Kuli-bang-bang, which jointly construct a diversion weir in the upstream portion of the Tina River. These *zanjeras* build a *palamag* (bundles of brush weighted down by stones) which is made of big branches of the *camachile* tree tied together. In interviews, *zanjera* members indicated that the design and technology used to build these structures were passed down from their ancestors who were founding members of the two *zanjeras*.

In another example, in Danum the materials and design of their dam are different from those of Pam-pan-niki and Kuli-bang-bang. Since the river current is deep and strong, the *zanjera* needs a sturdy base and firm structures for constructing its main diversion dam. Therefore, the members build many tepee-shaped structures made of bamboo poles, called *palomar*. These structures are placed in the water, filled with large stones, then covered with leaves and rice straw. These loose materials are anchored with more stones in and on top of each *palomar* so they will not float away. The *palomars* are linked together with other pieces of bamboo or wood.

One indicator of the practicality of the design of *zanjera* dams is that many of the dams have not changed much in appearance or in type of building material since they were originally constructed in the 1800s and 1900s. To ensure that someone is able to re-build the dams upon destruction by flooding, dam-building skills are shared by more than one member and passed across several generations of *zanjera* members. Given that the *zanjeras'* temporary dams wash out frequently, another advantage of the way the brush dams are designed is that the members can re-build them in different places in response to shifts in a river's course and volume, unlike permanent dams which silt up or are damaged when hit by a major typhoon. Members participate in making decisions about changing a dam's location when a shift in position is indicated by a change of the river course or other factors. *Zanjera* members meet to discuss where and how the dam should be re-built to meet changing conditions.

Operation and maintenance (repair)

To maintain their irrigation systems, *zanjera* members are involved in repairing or re-building brush dams, fixing damaged canals or structures such as intakes, and cleaning canals and farm ditches. These activities are performed on a regularly scheduled, routine basis as well as on an emergency, as-needed basis in response to changes in the environment. Routine repair activities include an annual, initial repair of brush dams, and semi-annual canal cleaning, checking and correcting of minor structural damages, plus other activities. Emergency or special repairs and maintenance occur when the brush diversions, culverts, access roads and other structures have been damaged due to typhoons, heavy rains and/or flooding.

Coward (1979) mentions two formats of labour required for repair, operation and maintenance activities in Danum. Both types of labour can be regularly scheduled or called for on an as-needed basis. The first type, *dagup* labour, requires the *zanjera's* entire membership for tasks such as major brush dam repairs or cleaning canals before the rainy season begins. The second means of organizing labour is *sarungkar*, for routine maintenance and repairs, which can be handled by smaller work parties, as well as for other functions.

PARTICIPATION IN INTERNAL ORGANIZATIONAL ACTIVITIES

Uphoff identifies decision-making, resource mobilization, communication and conflict management as four generic organizational activities in irrigation water management (Uphoff 1986). Two of these activities – *zanjera* resource mobilization and conflict management – especially contribute to *zanjera* management and environmental sustainability. One of the strongest characteristics of the *zanjeras* is the high level of participation by its members in organizational activities. *Zanjera* officers and members are known to work together closely.

Resource mobilization[25]

Construction and repair of *zanjera* dams, structures and canals require extensive labour, materials and food contributions and coordination of large work parties. Many of the *zanjeras* studied have detailed rules governing the mobilization of these contributions for *dagup* (major repairs and clean-up) work days. *Zanjeras* with *atars* favour proportional contributions. For example, when Danum holds a work day to repair its dam, work assignments are made both between and within *gunglos*. Work is assigned by *zanjera* officers in proportion to the number of *atars* in each *gunglo* and enforced by the *gunglo panglakayen*. On the other hand, to repair the brush dam, work is assigned to each *gunglo* by measuring and dividing the dam into equal portions. One section is assigned to each *gunglo* so there will be equal work for each group, regardless of the number of *atars* in each *gunglo*. Or, within each *gunglo*, when a *dagup* is scheduled, a notice instructs each member with one *atar* to bring a length of bamboo. At the *dagup* itself, some members are assigned to gather leaves or bundles of grass for the dam, while others have to gather and place stones on the diversion dam.

The repair and cleaning of brush dams and structures require a variety of materials to be contributed by *zanjera* members. Tools such as shovels, wheel barrows and *bolo* knifes are needed for digging and repairing the dams, and cutting overgrowth in the canals. Materials used to construct the diversion dams include stones, river rubble (gravel), leaves and brush, bamboo poles and wood. Rules exist which govern how these various resources are mobilized in most *zanjeras*. The kinds of materials and corresponding rules required for repairs and maintenance activities depend on the technical complexity and durability of the canals and structures, which in turn reflect the environmental conditions in which the irrigation system exists. The number of days worked in *dagup* labour shows a possible connection between work requirements and durability of dams. As of 24 October 1985 there had been over twenty *dagup* called for in *Zanjera* Danum, whereas in *Zanjeras* Pam-pan-niki and Kuli-bang-bang, there had been 12 or 13 *dagup*.

Conflict management and penalties

The longevity of many of the *zanjeras* can be attributed in part to their conflict management activities. Mediation is used by most *zanjera* officials to resolve conflicts within or between *zanjeras*. Even though some *zanjeras* have by-laws which mention the possibility of court action, most of those in the research area resolve conflicts internally, without resorting to litigation.

Member attendance at *zanjera* work activities is crucial to the operation and management of their irrigation systems. Fines *(multa)* are the most common penalty for breaking *zanjera* rules such as absence from required *zanjera* activities. Forty-one out of 45 of the *zanjeras* surveyed have fines for missing a *dagup* work day, while 20 out of 45 had fines for water distribution offences which include water stealing.

Given the scarcity and high value placed on irrigation water in Ilocos Norte, *zanjera* members consider water stealing as one of the most serious offences possible. This is reflected in the stiff penalties some *zanjeras* impose on anyone caught stealing water from a canal out of turn. In the majority of *zanjeras* surveyed, the fines for water stealing (a water distribution offence) are much higher than fines for other types of offences. Ten *zanjeras* specified a fine of P50 to P100 (US$ 2.50–5.00), which is a severe penalty for *zanjera* members.[26]

ZANJERA PARTICIPATION IN EXTERNAL ORGANIZATIONAL ACTIVITIES

External organizational activities are those related to parties outside an individual *zanjera*, whereas internal ones are within it.[27] *Zanjeras* have demonstrated their capacity to be involved with each other in irrigation management activities. Several large *zanjera* federations were observed in the research area, each of which had extensive organizational arrangements across two or more *zanjeras* to co-ordinate water allocation, distribution, resource mobilization and decision-making activities.

Several *zanjeras* have water-sharing arrangements with one or more others to allocate water throughout the wet and dry seasons. In one area of the municipality of Solsona there is a water-sharing committee which ensures that five shares of river water are allocated as agreed upon by the concerned *zanjeras*.

Mobilizing scarce resources and acquiring water were reasons one federation of two *zanjeras* formed in 1948 in the research area. One of them initially had its own brush dam which was downstream of the other. But in 1947 a devastating flood completely destroyed the downstream dam. The downstream *zanjera* found it extremely difficult to obtain water from the river after the flood despite its efforts to build a new dam. So it asked and was invited by the upstream one to join in a federation in 1948. Both *zanjeras* are now served by a single dam.

BENEFITS OF *ZANJERA* ACTIVITIES[28]

Rules and regulations guide the establishment of *zanjeras*, provide and protect members' access to scarce land and water resources, and govern activities such as system repair and maintenance, water allocation, meetings and social occasions. Benefits of the *zanjera* system include productivity and equity, which in turn contribute to the system's environmental sustainability.

Productivity

Zanjera farmers plant a variety of crops, including rice, corn, tobacco, mung bean, tomatoes and watermelon. However, rice is the main crop and the most preferred one of *zanjera* members. The *zanjera* irrigation system provides a more stable water supply to *zanjeras*. In 1985 rice production yield ranged from 1.5 to 2.6 tons/hectare for *zanjeras* located within the INIP project area (Sanyu Consultants 1986).

Zanjera members want to produce at least enough rice and other crops for their family's subsistence needs, reproduce inputs for the next season's crop(s) and, if possible, have additional produce to generate extra income to be used for their children's education, additional land purchases and other expenses. Some *zanjeras* surveyed have rules about cropping patterns and practices in the dry season.[29] These cropping rules govern which areas and what types of crops will be planted so that *zanjera* members have some assurance that they will be able to plant something during the dry season when water is scarce. A few of the *zanjeras* surveyed said they schedule a meeting near the end of the wet season in August or September to discuss and decide which areas will be planted with which crops and which areas, if any, should not be planted.[30] Knowledge about crops and planting conditions is pooled and traded during these discussions. After these meetings are held, members know what they may plant and plan accordingly.

An officer in *Zanjera* Danum discussed his cropping pattern. He plants three crops: first crop, rice; second crop, rice; third crop, mungo bean and corn. In this particular case the *gunglo* has the *atar* system with three different zones set up, so he can plant two different crops during the third crop period. He planted corn in the first and second zones and mung bean in the third zone in the *gunglo*. He explained that the *gunglo* meets and decides which crops should be grown in which zones so that scarce water can be allocated most efficiently for each zone. All farmers are expected to grow the same crop in the same zone, as agreed upon during the meeting. The officer also said that anyone who does not plant in an area which is scheduled for planting and water rotation (as a second or third crop) is subject to a fine (amount unspecified), so that valuable water is not wasted.

Equity

As noted earlier, the *atar* system of rights and responsibilities provides access to land and water for each member. Variations in access to water, location (head areas versus tail-end areas) and land productivity (soils, topography, etc.) are shared equally when each member has several parcels scattered both upstream and downstream. Unfortunately, as more *zanjera* members begin to hold only fractions of *atars* due to inheritance, this fragmentation reduces equity.

Members' obligations and responsibilities are assigned equitably in proportion to the number of *atars* cultivated. Obligations for *dagup*, routine cleaning and maintenance of canals, and material contributions for construction and maintenance are all proportional to the *atar* holdings, with equivalent amounts of labour and resource contributions required from each *atar*.

Environmental sustainability

Zanjeras are not immune to environmental problems associated with agriculture, such as soil erosion. Their proximity to steep mountains in the eastern portion of the province in combination with periodic, heavy rains during the typhoon season (May to October) subject the *zanjera* irrigation systems to flooding and subsequent soil erosion and sedimentation.

In Siy's study of a federation of nine *zanjeras* in Bacarra, Ilocos Norte, one of the reasons he gave for the development of the federation was 'the changeability of the river's course and characteristics which made the tasks of brush dam construction and repair increasingly difficult' (Siy 1982: 63). Two of the *zanjeras* in the federation are located on areas which were formerly parts of the river bed (Siy 1982). There are also examples of areas in Siy's study and in this research of *zanjeras* which lost part or all of their lands to erosion due to changes in a river's course.

Although *zanjera* members have no magic formula for controlling river channels, *zanjeras* are organized so that they can respond to erosion and some of the other problems in the canals. As previously mentioned, brush dams can be re-built in different locations in response to shifts in a river's course and volume, unlike permanent, concrete dams which silt up or are damaged when hit by a major typhoon.

Sedimentation of diversion structures and canals, and overgrowth of weeds in canals are common impediments to the operation of communal and bureaucratically run irrigation systems. *Zanjera* activities involve brush dam work and maintenance of canals which address these problems. Canals are cleaned of sedimentation and weed growth on a regular basis through the *sarungkar* work teams. When emergency flooding creates difficulties, such as, for instance, a blocked road culvert due to excess siltation, these same work teams can be notified by the call of a horn and respond quickly to remove the obstruction.

Emergency repairs by *zanjeras* are done on an 'as needed' basis, but procedures for scheduling these types of repairs are very specific, whether simple or complex, given the importance of the provision and timing of water for irrigated agriculture. Destroyed brush dams, intakes and major structures must be repaired as soon as possible to reinstate needed water supplies and/or to minimize flood damage to land and crops. The following example highlights the response of *Zanjera* Danum to a sedimentation problem in a 1986 typhoon, and shows the procedures for responding to two different kinds of emergency repairs. The initial emergency involved an immediate mobilization of labour, while the second, related problem required scheduling of a *dagup*.

In the first situation, while walking home from a visit to the *zanjera*'s main brush dam with the *mayor cabecilla*, the researcher and one of her assistants observed that the reinforced concrete (RC) pipe culvert in a canal passing under the road was clogged with debris, so that water was flooding the road and a nearby building. Since this occurred in the *gunglo* which is also the neighbourhood where the *mayor cabecilla* lives, he quickly went home to tell the farmers in his *gunglo* about the problem. One of these farmers blew the water buffalo's horn which is used to call *gunglo* members in case of such emergencies. Within 10 to 15 minutes the members arrived at the site with their hats, plastic sheets (their raincoats) and tools (*bolo* knifes and shovels), ready to work. But by the time the members arrived to clean the debris, the water level in the canal had greatly fallen and the water stopped overflowing onto the road. In just a few minutes the members cleared out the debris from the RC pipe and the canal water resumed flowing unobstructed. The immediate problem was solved.

But a second, more serious problem occurred; cleaning the debris from the RC pipe was not the end of this particular emergency. The significant reduction in the canal's flow of water may have provided temporary relief from local flooding, but it indicated a much bigger problem which became the concern of the entire *zanjera*: the reduced water level in the canal meant that part or all of the *zanjera*'s main brush dam had washed out and was no longer diverting water into the irrigation system's main canals and lateral. Thus, it was necessary for the general officers of the *zanjera* to schedule a *dagup*, or 'group work', to call all members to work on the following Thursday in order to re-build the dam.

CHALLENGES FACING *ZANJERAS*

The temporary nature of the *zanjera* brush dams is the greatest challenge to the *zanjeras*, according to some of the *zanjera* officers interviewed. Because the brush dams are made of non-permanent materials which are washed out easily, frequent repairs and maintenance are required. Some *zanjera* officers complained of the extensive work they have to do to organize and schedule voluntary labour by *zanjera* members for both routine and emergency repairs and cleaning of *zanjera* brush dams and canals.

Other *zanjeras*' officers complain that it is much harder now to motivate and coordinate members to contribute voluntary labour for irrigation tasks. A number of the *zanjeras* surveyed during the research loudly grumbled about members' absences on scheduled work days, and the fact that many of these absent members refused to pay (or simply ignored) the fines for absences.

Another difficulty facing *zanjeras* is getting access to additional sources of irrigation water. Many of them obtain water from individual river diversions or springs, acting independently of the others. In some areas obtaining water by individual *zanjeras* reinforces an 'upstream/downstream' pattern among them, where the downstream *zanjeras* feel that they are not getting enough water. It is difficult for downstream *zanjeras* to gain access to additional water sources. An inadequate water supply was the problem most frequently mentioned by *zanjeras* when asked in the research to list their most pressing problems.

In order to bring in 'additional irrigation water supplies', NIA designed and created INIP and its JICA-grant area (discussed below) to strengthen the 'inefficient', unlinked system of scattered *zanjera* brush dams. As seen in the following section, NIA and its INIP project introduced new and different requirements for

123

irrigation activities and for farmer involvement (or the lack of it) – NIA style, not *zanjera*-style – in irrigation development and organizational activities.

A NEW CHALLENGE TO THE *ZANJERAS*: INTERVENTION OF THE ILOCOS NORTE IRRIGATION PROJECT (INIP)

NIA introduced the Ilocos Norte Irrigation Project as an opportunity for improved irrigation development and increased agricultural productivity for the province. However, to some of the *zanjeras* the INIP was a challenge – to others a threat – to the existence and operation of their *zanjera* irrigation systems.[31]

In the early 1970s, two forces eventually led to the development of the INIP. First, both the provincial governor (Elizabeth Marcos Keon, sister of President Ferdinand Marcos) and the NIA Provincial Irrigation Office (PIO) recognized low agricultural productivity as one of Ilocos Norte's principal problems, and this was particularly aggravated by an inadequate irrigation water supply for the province. A second motivation for an irrigation project for Ilocos Norte was political. President Marcos made a personal request to NIA Central Office to investigate the possibilities of damming the Quiom-Maypalig River (Quiaoit River) in Ilocos Norte. Ostensibly this was to be for both flood control and irrigation purposes,[32] but a NIA staff member suggests that the real motivation for the project was to save Marcos's family home in Batac from the Quiom-Maypalig floods.

By 1975, extensive engineering and hydrological studies had been undertaken by NIA, and later by consultants, to explore possible dam sites. However, the growing recognition of the infeasibility of the Quiom-Maypalig Reservoir Project – due to surface material porosity and inadequate water supplies – led Governor Keon in 1977 to ask NIA Central Office to reactivate study of other possible irrigation water sources for Ilocos Norte, including the Palsiguan River.

Concurrent to Keon's pursuit of irrigation development for the province, the National Economic Development Authority (NEDA) was studying alternatives for a general rural development project for Ilocos Norte. NEDA – acting in its capacity as the Philippine national planning agency, and as the agency which oversees foreign-financed development projects like INIP – became further involved in activities with NIA, some Japanese consultants and the Japanese government which eventually led to INIP's creation.

In 1978 an area of eastern Ilocos Norte, with almost two hundred *zanjeras* operating in it, became part of the Ilocos Norte Irrigation Project (INIP) designed by NIA. NIA intended to introduce INIP as a new, national irrigation system to Ilocos Norte. INIP's total area was estimated at 22,600 hectares, or 76.2 square miles, almost twice the size of the city of Boston (Figure 5.2).[33] The plan aimed to increase agricultural production which would benefit 17,500 farm families in the area through the provision of improved irrigation facilities (JICA 1980b: 3). It also promised hydropower to provide electricity to Ilocos Norte. Plans for the project included two phases of development, called Phase I and Phase II; however, Phase II was officially shelved by NIA Central Office in December 1985. A 1,000-hectare pilot project, funded by the Japan International Cooperation Agency (JICA), and called the JICA-grant area, was contained within the boundaries of INIP's Phase I area.[34]

NIA applied a standard, engineering-oriented planning approach to the JICA-grant area and to preconstruction activities in its Phase I area. This standard planning technique focused on physical planning activities and ignored the social environment, including the *zanjeras*.

NIA ran into implementation and scheduling problems in the JICA-grant and Phase I areas. Farmers protested loudly against the project, and NIA's administrator from Manila, as well as some social scientists from Manila and Tokyo, came to see the project's problems for themselves. Based on their observations, NIA decided to undertake a more participatory approach as an alternative to the standard process. The revised approach promised opportunities for *zanjeras* to participate in INIP's ongoing planning activities.

JICA-GRANT PROJECT PLAN

NIA requested a Japanese survey team to develop the JICA-grant project. The JICA-grant project was designed to provide irrigation and drainage systems with 'terminal facilities' to an upstream area served by the Labugaon River in the municipality of Solsona. This project was to assist farmers with water management techniques of terminal facilities. The Japanese team focused on these techniques because they said that it is the: 'terminal water management technique which is a bottleneck to increase [sic] the effect up to the

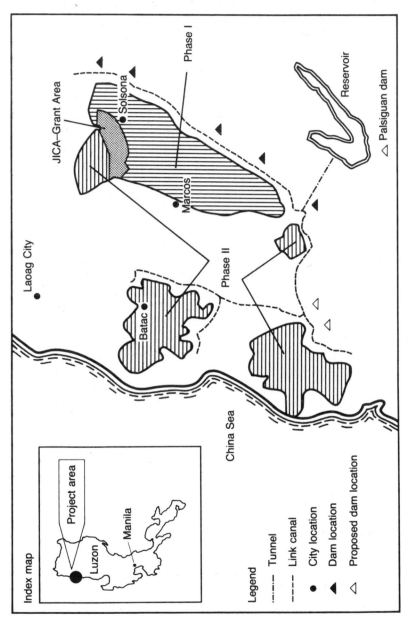

Figure 5.2 Ilocos Norte Irrigation Project (INIP): regional locator map

Source: Yabes 1990

target in many irrigation projects which have been implemented' (JICA 1980a: 24). The JICA-grant project was to be finished prior to construction of Phase I components (JICA 1980a: 6, 25).

INIP'S PHASE I PLAN

In Phase I, INIP project planners proposed to build a new network of irrigation and drainage canals supplied by five new major diversion dams. These five dams were to be constructed across the Labugaon, Solsona, Madongan, Papa and Nueva Era Rivers. These new irrigation networks proposed to cover an area including almost two hundred *zanjeras*, though no details were given in INIP's feasibility study of ways of integrating these existing *zanjera* systems into the new NIA network of irrigation canals and dams.

INIP's plan acknowledged the presence of communal irrigation systems in the area, but gave them very little attention within its engineering, agricultural or institutional components. The existing systems were mentioned only 11 times – for a total of less than 3 pages – in a 200-page report! Furthermore, they were never called by their correct name of '*zanjeras*'. The feasibility study cited water losses due to deteriorated irrigation and drainage systems having temporary brush dams, and 'insufficiency in water management due to intricated [sic] canal networks' (JICA 1980b: 4–1). Thus, the plan proposed to modernize agricultural production, where: 'In the Project Area, irrigation and drainage canals will be newly constructed and/or improved to separate irrigation and drainage systems, and terminal facilities will be rearranged and reinforced' (JICA 1980b: 4–3).

Planning and construction activities of the JICA-grant project proceeded at a rapid pace in early 1981. NIA used a standard planning process characterized by little interaction with users and little attention to the existing physical or organizational structures of the *zanjera* irrigation systems to design the JICA-grant area. Problems occurred in the JICA-grant and the Phase I areas.[35]

ZANJERA RESISTANCE TO THE JICA-GRANT PROJECT

The original design of the JICA-grant area was eventually changed, due to resistance by three upstream *zanjeras* which refused to join the project. Trees were felled to block the passage of construction equipment. Farmers with *bolo* knives threatened NIA staff who tried to enter the project area. During interviews members and officers of these non-joining *zanjeras* listed several reasons for resisting the project, including the fact that they had enough water without the NIA project. Others said they did not want to pay irrigation fees to NIA. The *zanjeras* doubted NIA could provide any extra water and were unwilling to have to pay fees for water they had always received for free or for a much lower water rights fee. Another reason some resisted the project was that they were afraid many of their small farmlands would be erased or greatly reduced by the wide canals, access roads and facilities proposed by NIA. They also said they wanted to keep their existing *zanjera* irrigation system and organization which had operated for years, but which would be eliminated and replaced by NIA's new irrigation system.

The upstream *zanjeras* also were against the NIA project because they did not want to give up their water rights or grant NIA the right-of-way (ROW). One major concern for *zanjera* members in the JICA-grant area was that some of the new canals might take up large portions of their small plots. Since it was the landowners, and not the tenants, who were to be compensated, the latter also feared that the source of their livelihood would be reduced with nothing received in return.

In the face of strong opposition from these *zanjeras* in the upstream portion of the JICA-grant area, NIA changed the location of the area served by the JICA-grant project. The NIA Central Office decided to exclude approximately 500 hectares in the upstream area of the original JICA-grant plan where the *zanjeras* refused to join, and include an additional 500-hectare downstream area where *zanjeras* were willing to be part of the JICA-grant project.

In NIA's effort to solve problems in the upstream portion of original JICA-grant area, a memorandum of agreement was signed on 28 March 1981 between NIA and the federation of the three *zanjera* holdouts through which the latter:

> agreed to let their brush dam and intake canal to be improved by the [INIP], which will be used by both the *zanjeras* and NIA, while another NIA permanent dam is being constructed upstream of Labugaon River. They also agreed to let the main canal of the NIA traverse their *zanjeras* as long as NIA will shoulder all expenses in constructing facilities and structures

in the intersections between their canals and those of the NIA. Incorporated also in the agreement is the provision for the continuous release of water by the NIA's permanent dam to the dam of the *zanjeras* in accordance with their water rights when the permanent dam of the NIA will already be put to use.

(Padanum 1981: 8)

Despite these design changes in the JICA-grant project's location, other system design issues – personnel, design criteria and opportunity for user input – remained the same. NIA personnel responsible for the JICA-grant project's design were predominantly civil and agricultural engineers who emphasized modern technology in canal and structure designs. For example, NIA's standard design criteria were used to determine the number and location of turnouts (one turnout per rotational area of 30 hectares) (JICA 1980a). This was in contrast to *zanjera* turnouts which served varying areas ranging from 10 to 40 hectares.

A revised, participatory planning approach to INIP

Eventually there was a transition to a revised, more participatory planning approach in the Phase I area activities for INIP, beginning in October 1981. In light of *zanjera* farmer protests against the project, and based on observations during several visits to the JICA-grant site, NIA administrator Fiorello Estuar reviewed the JICA-grant project and INIP plans and made the very unusual and drastic decision to drop the standard designs. Estuar ordered the Phase I area of INIP to be resurveyed and redesigned according to a revised planning approach.

INIP's objectives were modified, emphasizing rehabilitation over new construction, and recognizing the importance of a role for the *zanjeras* in project activities. Four guidelines for the revised approach were recommended by the social scientist team and endorsed by the administrator and the NIA Central and project field offices (Visaya 1982: 4):

1 Preserve the identity of the *zanjera* groups.
2 Follow existing canal lines as much as possible.
3 Conceive the project as rehabilitation of existing communal irrigation systems, not as construction of a new, large-scale system.
4 Involve farmers in planning and implementing the project.

129

INIP's participatory approach was achieved through organizational changes in NIA, NIA-*zanjera* communication-facilitating workshops, and the inclusion of third-party researchers and academics to provide process documentation for NIA's decision-making. This resulted in more interaction between NIA and the *zanjeras*, with greater attention being paid, in the process of project design, to the latter's physical and organizational structures, and to their special development needs.

On the contra side, the participatory approach was resisted by some INIP staff because it was felt to be too time-consuming, and because it would increase the complexity of project tasks, raise project costs, expand staff accountability, and change the status quo of planning methods.[36]

INIP'S ENVIRONMENTAL CONSEQUENCES FOR THE *ZANJERAS*

INIP has aggravated existing environmental difficulties, such as flooding, for the *zanjeras* in certain areas of the project; it has also caused and contributed to new problems – soil erosion and sedimentation – in other project locations.

Flooding

Flooding is an annual problem of the typhoon season in eastern Ilocos Norte which occurred long before the creation of INIP. However, INIP's arrival to the area has intensified flood damage in several locations scattered throughout the project. For example, a September 1985 flood occurred in Nueva Era. Affected landowners and tenants, including the mayor, were very angry at NIA because their fields were damaged or ruined by siltation from NIA's irrigation system. According to Nueva Era *zanjera* farmers this type of flood damage had never occurred in this particular area prior to the construction of the NIA system.

In another case, severe flooding in 1985 and 1986 in the Madongan Left irrigation system of INIP destroyed a number of NIA irrigation structures and damaged portions of the main canal and one of the lateral canals. Damage was so severe that the NIA system was inoperable and consequently abandoned in several locations. Some of the *zanjeras* were unable to resume operation in areas adjacent to the flood damage.

130

Throughout INIP design and construction activities, NIA staff have tried to address the severe flooding problems in the region by incorporating flood control walls and structures within the irrigation project. However, because inclusion of flood control structures would significantly raise project costs they were not included in the original project plans. Gaining approval for the structures has been a very slow and somewhat unsuccessful venture for NIA staff. As of 1986 one major flood control wall was constructed in the Papa system of INIP; no evaluation had been made as of that time as to how effectively the wall had controlled flooding.

Soil erosion and siltation

INIP activities and structures have introduced new soil erosion and siltation difficulties for the *zanjeras*. For example, in the Nueva Era irrigation system one of the newly constructed lateral canals ran uphill, and, due to poor compaction of one of NIA's link roads, severe soil erosion had buried fruit trees and rice fields because of flooding. Also, in the same area, the contractor dumped construction materials on land not included in the right-of-way negotiations, and some farmland was destroyed due to flooding and siltation. At a *zanjera*'s request, the Japanese consultant representative visited the site at least three times to see the uphill canals, erosion and road's poor compaction. *Zanjera* members and affected residents were very angry when neither the Japanese consultant nor NIA took any action against the offending contractor; soil erosion continued to bury the nearby rice fields. Nor did NIA or the consultant try to correct the soil erosion problems during 1985 or 1986. Despite repeated *zanjera* complaints about the lateral canal that allegedly ran uphill, no action was taken by NIA or the contractor to correct this problem. Eventually NIA dropped this lateral from the project saying there would not be enough water from the dam to serve it. However, a trail of soil erosion from canal construction activities was left in the wake of the abandoned lateral canal.

The revised, participatory planning approach introduced mechanisms through which *zanjeras* could register complaints with NIA, such as forwarding complaints through agri-institutional workers, and direct meetings with INIP management. However, despite the greater opportunity for voicing complaints in the revised approach, the act of complaining did not help prevent these problems from occurring, or bring any acceptable solution to the concerned *zanjeras*. The *zanjeras* did not attempt to correct the siltation problems themselves, viewing the problem as within NIA's domain. Thus, the affected fields remained unproductive and buried by silt.

Two of INIP's dams were regularly plagued with substantial upstream soil erosion and subsequent sedimentation of the diversion dams and intake structures. At the Solsona Dam, upstream soil erosion was apparently caused in some or large part by deforestation and inadequate reforestation of the upstream water-shed areas which supplied timber to the dendro thermal plant located to the north side of the dam.[37] As a result of this upstream soil erosion run-off, the Solsona Dam filled up with gravel and rock sedimentation which clogged the dam and the intake canal which served the entire Solsona Right irrigation system in INIP. NIA equipment was used to clean out the sedimentation without *zanjera* assistance.

In the second case of the Nueva Era dam, excessive upstream soil erosion – possibly caused by past illegal mining activities – entirely buried the dam at least twice as of 1985. NIA eventually used its construction equipment to uncover the dam and de-silt the intakes. Nonetheless, the uncovering process took a long time, so meanwhile the affected *zanjeras* re-built their brush dams and continued to operate their own *zanjera* irrigation systems in the interim period while awaiting repair of the NIA dam.

Environmental consequences of INIP for the *zanjeras*

A major drawback of INIP for the environmental sustainability of the *zanjeras* was that it apparently weakened the voluntary labour system for repair, maintenance and operation activities in some of the *zanjeras*. In contrast to the *zanjeras'* use of *voluntary* labour only for irrigation system maintenance and management activities, NIA tried to take some well-meaning shortcuts by assisting in the repair of *zanjera* brush dams, by using NIA construction equipment for spontaneous, quick repairs, thinking that this would help save the *zanjeras* time and hard labour. Unfortunately there were a few unforeseen consequences of this equipment assistance. For one thing, when NIA did the work for the *zanjeras*, some of the *zanjera* members refused later to clean canals or repair their brush dams, with the excuse that NIA equipment could do it faster and easier than they could do it manually. Also, the attitude of some absent members was that it was NIA's project, so NIA should repair the dams and clean the canals which were being disturbed by NIA's activities. The assistance of the equipment operators later created problems for some of the *zanjera* officers, who had great difficulty getting their members to clean canals when NIA was not present or available to help.

From another perspective, the physical size and scope of the dams and some of INIP's structures are beyond the manual capacity of the *zanjera*'s organization. It is difficult and oftentimes impossible for *zanjera* members to remove heavy sedimentation from INIP's dams and intake structures. Although NIA and the *zanjeras* mutually agreed that upon project completion, operation and maintenance of INIP's dams and head intake structures would be NIA's responsibility, this agreement in effect created another challenge to the long-term environmental sustainability of the irrigation system. Based on this agreement, *zanjeras* are now dependent on NIA equipment and staff to make major repairs to the dams and head intakes. Sometimes, during partial operation of the INIP irrigation system, NIA was observed to respond quickly to damaged or silted dams and structures. However, on several occasions it took NIA months to repair major damage, due to lack of funds and materials, and to bad weather. These delays on repair and clean-up activities by NIA interfered with the delivery of irrigation water in a timely fashion to the *zanjeras*. These delays occurred even when the project was in the midst of a full construction schedule with available funds to make the repairs. *Zanjeras* and INIP project staff expect even greater delays in repairs and major clean-ups after the project is completed, when project funds will dry up.

LESSONS FOR ENVIRONMENTAL SUSTAINABILITY: ISSUES IN PARTICIPATORY PLANNING APPROACHES TO PROJECT INTERVENTION IN AREAS WITH TRADITIONAL RESOURCE MANAGEMENT SYSTEMS

The INIP project represents a case illuminating the potential for applying participatory elements in order to re-structure a large national irrigation activity where local irrigation systems already exist. It yields several lessons about what might be emphasized or avoided when implementing participatory planning approaches to promote environmental and agricultural sustainability in traditional resource management systems in the Philippines and in general.

In the INIP project area, one found a whole set of active irrigation groups – the *zanjeras* – which fit within a category NIA knows about (communal irrigation groups), and around which the NIA agency has organized a Communal Irrigation Program (de los Reyes and Jopillo 1986; Korten and Siy 1988; Korten 1988; Jopillo and de los Reyes 1988). *Zanjera* irrigation systems are built, operated and maintained by the users of the water. They perform various irrigation tasks and manage conflict. Despite this agency knowledge of communals, one crucial issue in INIP's project planning which must be emphasized is that INIP was conceptualized as a 'new' national irrigation project rather than as rehabilitation project of a collection of existing communal irrigation systems.

NIA's 'participatory programme' for communal irrigation systems, begun in 1976, includes elements of local resource mobilization and establishment of local organization forms. These elements of irrigation development were already known to NIA and available in the *zanjeras* when NIA and INIP arrived. Once one understands the impressive organizational capacity of the *zanjeras* described in this chapter, one can anticipate the *zanjeras'* resistance to the imposition of the INIP project.

Had NIA and the Japanese consultants chosen to develop INIP as a collection of communals, presumably all of NIA's experiences with participatory approaches to communal irrigation systems might have been mobilized. The apparent lack of debate among INIP planners about the initial conceptualization of project type (national versus communal) was a serious oversight.

This oversight and the problems of the INIP irrigation project – e.g. flooding, soil erosion, sedimentation – can be corrected. NIA already has begun to take steps in that direction. As previously stated, the thrust of INIP's revised, participatory planning approach is to preserve the integrity of the *zanjeras'* organization by designing the irrigation project in a way which builds on the strengths of the original character of the traditional resource management systems rather than trying to change completely those traditional systems.

The irrigator association's sense of ownership in the national system should also be encouraged through a new water rights policy which allows any existing communal irrigation group to retain water rights when their systems are absorbed. (This is in contrast to the initial INIP water rights policy which absorbed all *zanjera* water rights.) The new policy should also give direct water rights to any new irrigator associations which register with securities agencies like the Securities Exchange Commission (SEC) in the Philippines, and which are created as part of the national project.

The time frame for projects, large and small, should include a sufficient period for community organizers to contact community leaders and organize irrigator associations. The organizers should be sent out several months prior to project formulation and negotiation by potential project donors or contractors. Except in cases of political or personal danger, these organizers should live in their assigned areas.

New and expanded measures of project progress and evaluation should be created. The institutional activities of agency staff such as Agricultural Coordination Division (ACD) should be included in project progress reports in association with detailed coverage of physical accomplishments. Genuine opportunities should be provided by NIA for farmers to raise legitimate concerns and complaints, and to allow NIA to make timely and adequate responses to these concerns.

Finally, a regional perspective and multi-sector environmental analysis and approach is essential for planning and implementing large irrigation infrastructure projects, especially when those projects encompass existing traditional resource management systems. In this particular case, not only the irrigation sector should be considered, but also the potential effects of other factors such as floods – which require flood control mechanisms – and water-sheds – the stripping of which lead to water and soil run-off.

The INIP case has some limitations as a general case of implementing a participatory planning approach. As we have seen, the *zanjeras* are a very specific kind of communal group, with complex participatory practices which are not always present in other irrigation groups, or other kinds of local organizations. In a future case of participatory planning, it would be wrong for project planners to assume that existing local organizations might have the same kinds of participatory practices and capacities as those of the *zanjeras*. Thus, a key initial part of any participatory planning approach to promote environmental sustainability is to profile the characteristics, history, context, capabilities and weaknesses of the potential local organization before any project decisions are made.

NOTES

1 Ruth Ammerman Yabes is with the Department of Planning, Arizona State University, Tempe, Arizona. This paper is an amended version of Yabes (forthcoming) and Yabes (1990: 11–51). The author wishes to thank Robert Yabes, Porus Olpadwala, E. Walter Coward, Norman Uphoff, Susan Thompson and Johanna Looye for comments on earlier drafts of this paper. Research was conducted in Ilocos Norte, Philippines, in 1985–6 with Fulbright-Hays and National Science Foundation doctoral dissertation grants.

2 Christie 1914; Coward and Siy 1983; Coward 1979; Lewis 1971, n.d.; Siy 1982; Thomas 1978; Visaya 1982.

3 This number is the 1988 estimate by the Provincial Irrigation Office (PIO) of the National Irrigation Administration (NIA). No total population figure is known of the number of persons involved with these systems. A discussion of the different estimates of the number of *zanjeras* is contained in Yabes (1990), Appendix E.

4 Christie 1914: 99. Lewis (1971: 130) cites 17,000 hectares in 1971. Thomas (1978) provides data obtained from the Provincial Irrigation Office of the National Irrigation Administration in Laoag City.

5 Keesing (1962: 27) cites references to irrigation canals by Juan de Medina (1630) and later comments (late nineteenth century) by editor Father Coco in Blair and Robertson 1903–9, vol. 13: 247, 276–9; vol. 7: 174; vol. 12: 210.

6 Farmer survey, Yabes field research 1986. Descriptive statistics in the following sections are summarized from results from *zanjera* and farmer key informant instruments administered during field research activities in Ilocos Norte in 1986 (Yabes 1990).

7 Coward and Siy (1983: 6) discuss three types of *zanjeras*, depending upon the status of the land at the time of inception (whether it was cultivated; or uncultivated) and upon the type of labour used to construct them (whether they were built by landowners, non-landowners [tenants or paid labour], or both landowners and others). See also Yabes 1990: 14–16.

8 Siy was the first author who wrote about the *biang ti daga* agreements (1982).

9 Though prevalent in many of the *zanjeras*, not all *zanjeras* are based on the *atar* system.

10 See Coward 1979: 29–30; Siy 1982: 32–6; Coward and Siy 1983: 3–8; and Lewis n.d. 53–7; for their discussion of the *atar* share and the arrangement of *zanjera* landholdings.

11 Yabes (1990), Appendix G, carries a detailed description of the *atar*.

12 An extensive description of the physical layout and organization of this *zanjera*, in Dingras, Ilocos Norte, is provided by Coward (1979: 28–36). Note that pseudonyms have been used for all *zanjeras* named in this chapter to protect the confidentiality of information provided by their members.

13 The term *dagup* is used by *zanjeras* throughout the research area to describe work days when the entire membership of a *zanjera* is expected

to work. *Dagup* labour requires all members to make labour contributions for major repairs to the main brush dam, or to clean long sections of the main canal and laterals.

14 It may be surprising, but the presence of cooks is essential to larger *zanjeras*.

15 Coward does not make this distinction between the landowner's *zanjera* officers and the *zanjera* members as officers who actually cultivate the agricultural lands and operate the irrigation system (1979: 31–2). For additional information about *Zanjera* Danum, see Viernes, 1986.

16 The following analysis is adapted from discussion of the question, 'Who participates?' in Cohen and Uphoff 1977.

17 Lewis n.d.: 48–60. See Appendix J (Yabes 1990) for further details about *zanjera* membership and non-membership categories.

18 In the case of larger *zanjeras* which have *gunglos*, the *gunglo* is divided into zones instead of the *zanjera*.

19 1903 Census of the Philippines, in Siy 1982: 20.

20 Appendix H (Yabes 1990) provides an extended discussion of *zanjera* rules and regulations.

21 These are not exclusive categories. Uphoff discusses the interrelation among all of these irrigation activities, as highlighted in his three-dimensional matrix of irrigation management activities (Figure 1 in Uphoff 1986: 42). The three dimensions in the matrix encompass water use, control structure and organizational activities.

22 Much of this discussion is taken from Yabes (1990: 20–40).

23 Interviews, Yabes field research, 1986.

24 *Zanjera* survey, Yabes field research, 1986.

25 See Siy (1982) and Yabes (1990: 291–2) for further details about *zanjera* resource mobilization.

26 As of 1975, the annual per capita income in this region was only 780 pesos, compared to the national average per capita income of 895 pesos (JICA 1980b: 1).

27 Uphoff regards decision-making, resource mobilization, communication and conflict management as categories of external as well as internal organizational activities (1986).

28 Much of this section is taken directly from Yabes (forthcoming) which was written in response to an outline and suggestions made by Wilbert Gooneratne.

29 Not all *zanjeras* have cropping rules, including those which plant only one irrigated crop.

30 Most of the *zanjeras* surveyed listed their cropping patterns for one, two and sometimes three crops. Several of the *zanjeras* have water available only for one rice crop and plant nothing else. Other *zanjeras* first plant rice, and then plant a few vegetables even though no irrigation water is available for a second crop.

31 Yabes (1990: 68–83), Chapters 4, 5, 7 and 8 and Appendices K, L and M, provide a detailed description of INIP – its origins, components, obstacles and opportunities. See also Angeles *et al.* 1986.

32 On 31 January 1973 President Marcos sent the following directive to NIA Administrator Alfredo Juinio: 'Have the possible construction of a dam in the Kiom-Maypalig [sic] River of Batac, Ilocos Norte studied for both flood control and irrigation. Send me a report on this' (Marcos 1973).

33 Prior to a workshop held at NIA Central Office in December 1985, INIP was known as the Palsiguan River Multipurpose Project (PRMP), named after the Palsiguan River in Abra which was included in Phase II of the project. Following recognition during the NIA workshop that Phase II would not be implemented, the PRMP project came to be referred to as 'INIP'.

34 The 'pilot project' is the local name for the Terminal Facilities project. The 'pilot' project was neither innovative nor precedent in terms of time or space. However, the project did serve the purpose of a pilot in terms of showing NIA planners what not to do during subsequent project activities in Phase I of INIP. But the 'pilot project' will be called the 'JICA-grant area' in this chapter to avoid confusion with the general meaning of the term 'pilot'.

35 These problems are discussed in Yabes (1990: 105–10) and in Siy (1987).

36 Yabes (1990: 219–27) includes an extended discussion of the difficulties in the transition to INIP's revised, participatory approach.

37 Verification of the sources of soil erosion in this and the next case was impossible since the upstream areas of both dams were located in areas set off-limits by the Philippine Army due to military activities by the New People's Army (NPA).

REFERENCES

Angeles, H., Saplaco, R., Agbuya, M.C., Sicat, E., Calanoc, N., Tejada, G. and Alipio, R. (1986) 'Process documentation research on the development/rehabilitation of communal systems in the NIA-Ilocos Norte Irrigation Project', ISMIP paper no. 1, Muñoz, Nueva Ecija, Philippines: Central Luzon State University.

Barker, Randolph, et al. (1984) 'Irrigation development in Asia: Past trends and future directions', Ithaca, New York: Cornell Studies in Irrigation, Report 1.

Blair, Emma H. and Robertson, James A. (eds) (1903–9) The Philippine Islands, 1493–1803. Cleveland: A.H. Clark.

Christie, Emerson B. (1914) 'Notes on irrigation and cooperative irrigation societies in Ilocos Norte', The Philippine Journal of Science, 9 (2): 88–113.

Cohen, John M. and Uphoff, Norman T. (1977) Rural Development Participation: concepts and measures for project design, implementation and evaluation, Ithaca, New York: Rural Development Committee, Cornell University.

Coward, E. Walter, Jr (1979) 'Principles of social organization in an indigenous irrigation system', Human Organization 38 (1): 28–36.

Coward, E. Walter, Jr (1980) *Irrigation and Agricultural Development in Asia: perspectives from the social sciences*, Ithaca, New York: Cornell University Press.

Coward, E. Walter, Jr, and Siy, Robert Y. Jr (1983) 'Structuring collective action: an irrigation federation in the northern Philippines', *Philippine Sociological Review*, 31: 1–2, 3–17.

de los Reyes, Romana P. and Jopillo, Sylvia Ma. G. (1986) *An Evaluation of the Philippine Participatory Communal Irrigation Programme*, Quezon City, Philippines: Institute of Philippine Culture, Ateneo de Manila University.

Eggink, Jan W. and Ubels, Jan (1984) 'Irrigation, peasants and development', M.Sc. thesis, Agricultural University of Wageningen, Netherlands.

JICA (Japan International Cooperation Agency) (1980a) 'Basic design report on Terminal Facilities Project in Ilocos Norte Irrigation Project', n.p. August.

JICA (Japan International Cooperation Agency) (1980b) 'Overall plan for Ilocos Norte Irrigation Project in the Philippines', n.p. December.

Jopillo, Sylvia Ma. G. and de los Reyes, Romana P. (1988) *Partnership in Irrigation: farmers and government in agency-managed systems*, Quezon City, Philippines: Institute of Philippine Culture, Ateneo de Manila University.

Keesing, Felix M. (1962) *The Ethnohistory of Northern Luzon*, Stanford, California: Stanford University Press.

Korten, Frances F. (1988) 'The working group as a catalyst for organizational change', in Frances Korten and Robert Siy, Jr (eds) *Transforming a Bureaucracy: the experience of the Philippine National Irrigation Administration* 61–89, Connecticut: Kumarian Press.

Korten, Frances F. and Siy, Robert Y. Jr (eds) (1988) *Transforming a Bureaucracy: the experience of the Philippine National Irrigation Administration*, West Hartford, Connecticut: Kumarian Press.

Lewis, Henry T. (1971) *Ilocano Rice Farmers: a comparative study of two Philippine barrios*, Honolulu: University of Hawaii Press.

Lewis, Henry T. (n.d.) *Ilocano Irrigation: corporate groups and oriental nepotism*, Honolulu: University of Hawaii Press, TMs, photocopy, unpublished book manuscript.

Marcos, Ferdinand. (1973) Note to Alfredo Juinio, 31 January, photocopy.

Ministry of Agrarian Reform (MAR) Team Office, Dingras, Ilocos Norte (1984) in agricultural progress report: Palsiguan River Multi-Purpose Project (Phase I area) (Loan No. PH-P32 and PH 45), report 4, National Irrigation Administration, Ilocos Norte, Philippines, 31 December, mimeo.

Padanum (1981) 'ROW problems at the Pilot Project settled', 2 (3): 8.

Sanyu Consultants, Inc. (1986) 'The five-year agricultural development plan of the Ilocos Norte Irrigation Project (INIP): (Stage I)', Ilocos Norte, Philippines, photocopy.

Siy, Robert Y. Jr. (1982) *Community Resource Management: lessons from the zanjeras*, Quezon City, Philippines: University of the Philippines Press.

Siy, Robert Y. Jr. (1987) 'Averting the bureaucratization of a community-managed resource: the case of the *zanjeras*, in David Korten (ed.)

Community Management: Asian experience and perspectives, West Hartford, Connecticut: Kumarian Press.

Tamaki, Akira (1977) *The Development Theory of Irrigation Agriculture*, Tokyo: Institute of Developing Economics, special paper 7.

Thomas, William L. (1978) '*Zanjeras*, communal irrigation systems in the province of Ilocos Norte, Philippines: a corrective note and up-dating', TD, photocopy, unpublished manuscript article, January.

Uphoff, Norman (1986) *Improving International Irrigation Management with Farmer Participation: getting the process right*, Boulder, Colorado: Westview Press.

Viernes, Villamor (1986) 'From the *zanjeras*: another priceless heritage', *Pangawidan ti Amianan*, 10 (1): 4–5, 28.

Visaya, Benito P. (1982) 'The Palsiguan River Multi-Purpose Project and the *zanjeras*', paper for conference on 'Organization as a strategic resource in irrigation development', Asian Institute of Management, Manila, 15–19 November.

Yabes, Ruth Ammerman (1990) 'Obstacles and opportunities of participatory planning in a large irrigation system: the case of the Ilocos Norte Irrigation Project (INIP) in the Philippines', unpublished Ph.D. dissertation, Cornell University.

Yabes, Ruth Ammerman (forthcoming) '*Zanjeras*: Philippine communal irrigation systems in the north', in Wilbert Gooneratne *et al.* (eds) *Traditional Community Irrigation in Asia: a survey of customary rules and regulations*, Nagoya, Japan: United Nations Centre for Regional Development.

Yoder, Robert (1986) 'The performance of farmer-managed irrigation systems in the hills of Nepal', unpublished Ph.D. dissertation, Cornell University.

6

SUSTAINABLE DEVELOPMENT AND PEOPLE'S PARTICIPATION IN WETLAND ECOSYSTEM CONSERVATION IN BRAZIL:[1] TWO COMPARATIVE STUDIES

Antonio Carlos S. Diegues

Traditional communities in Brazil provide many examples of community-level natural resource management which is ecologically sound and economically feasible. These management systems reflect an essential pattern in the relationship between traditional human communities and their environment. Many of the techniques and approaches used are based on a deep knowledge of the environment, and on a tradition in which community livelihood derives from the sustainable use of natural resources.

Most literature on people's participation, however, speaks of a need to involve local populations in development or conservation projects which are frequently designed by outsiders (who may be consultants, technical experts or government departments), and which aim to improve the living conditions or environmental situation of local populations through technological or socio-economic changes. These projects are based on the assumption that the so-called 'traditional' societies are necessarily static and need to undergo changes originating from outside in order to achieve 'development'.

Such 'development' is handled by the 'modern' local elite who, as a rule, take advantage of the resulting economic growth. The resulting 'modernization' usually carries no benefits for the local population. In this context, it is often assumed that peasant communities are opposed to change because of their traditional values.

Studies in Brazil (Diegues 1983; Forman 1970) and in other parts of the world, however, have shown that peasants are open to change provided that they benefit from the outcome of the process, and that the risks involved in adopting innovations do not threaten their livelihood.

According to the 'modernization' theory, traditional management techniques are primitive and ineffective and are in need of being replaced by 'modern' ones, although the traditional techniques result from a long process of ecological and cultural adaptation and respond to a situation where demand for renewable resources is not high. Once demand is increased, consumption may rise correspondingly, leading to the overexploitation of natural resources. As a result, the traditional and previously efficient management techniques are abandoned by the local population (Bourgoignie 1972).

In this chapter, the concept of 'traditional communities' relates to a type of economic and social organization with little or no capital accumulation and no demand for hired labour, and in which a society of independent producers is involved in small-scale activities such as agriculture and fishing, fruit harvesting and production of handicrafts. Economically, therefore, these communities are based on the intensive use of renewable natural resources. An important feature of this 'petty' production is the producers' knowledge of natural resources, biological cycles, and of different species and their eating habits.

This traditional know-how, passed from generation to generation, is an important tool in the conservation of natural resources. As these producers usually have no other source of income, the sustainable use of natural resources is of fundamental importance. Their consumption patterns, low population density and low level of technological development lead to very limited interference with natural processes and to the conservation of natural ecosystems. Important characteristics of many traditional societies include the combining of several economic activities, the recycling of wastes and a relatively low level of pollution. Conservation of natural resources is also an integral part of the culture, an idea expressed in Brazil by the word *respeito*, meaning respect, and used not only vis-à-vis nature but also vis-à-vis other members of the community.

However, this small-scale commodity sector no longer operates in isolation, and is now frequently found in combination with capitalist structures. When the capitalist mode of production is introduced, important social and economic changes occur which deeply affect petty production and the relationship between humans and nature. There is very often a progressive disruption of the cultural patterns responsible for the conservation features within traditional societies. Small-scale producers, becoming dependent on cash, are forced to intensify their exploitation of natural resources, and they thus bring about environmental degradation. Increased productivity does not necessarily lead to better income or living conditions for the majority, since the profit ends up in the hands of a few, such as traders or government officials.

This chapter presents two case studies which highlight the relationship between community participation and conservation of the natural environment. In each case the local community takes action after losing land as a result of expanding capitalist production in regions rich in natural resources. This expansion leads not only to serious degradation of the environment and its natural resources, but also threatens the livelihood of the traditional population, disrupting indigenous production practices and turning small producers into wage-earners.

The first study concerns the Marituba floodplain of the São Francisco River, in the north east of Brazil, and the opposition of the local population to its proposed conversion into an irrigated area for intensive rice culture (see Figure 6.1). The rivers and lakes are 'commons', but the land has been progressively taken over by capitalist sugarcane plantations. To guarantee their livelihood, the fishermen and peasants have fought for the creation of a special environmental protection area.

In the second study, the rubber tappers of the Guaporé Valley, in the state of Rondonia (see Figure 6.1), have been deprived of access to rubber trees as the land is sold to rich capitalists from the south who burn the forest for logging and cattle rearing. The rubber tappers, or *seringueiros*, oppose the destruction of the forest which represents the source of their livelihood.

CASE STUDY 1: PEOPLE'S PARTICIPATION IN THE CONSERVATION OF THE MARITUBA WETLANDS

The first case study looks at the resistance of fishermen and peasants to the transformation of the last large floodplain of the São Francisco River into an irrigated area for intensive rice culture (see

143

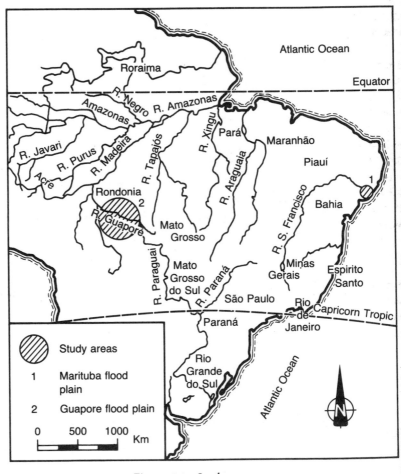

Figure 6.1 Study areas
Source: Simielli Maria Helena (1988) *Geoatlas*, São Paulo: S.P. Ed. Atica.

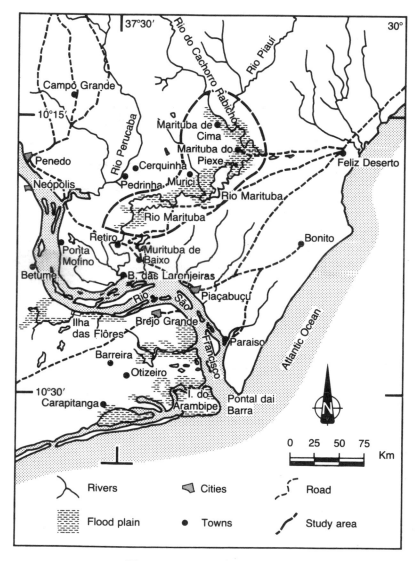

Figure 6.2 Marituba floodplain

Source: Fundacão Instituto Brasileiro de Geografia e Estatistica (IBGE) (1976) *Folha Aracaiu*, Rio de Janeiro: IBGE.

Figure 6.2). Marituba is a floodplain, or *várzea*, lying to the east of the São Francisco River, in the state of Alagoas, in the north east of Brazil. It covers approximately 200 square kilometres of marshland, formed by the periodic flooding of the river. The marsh is crossed by the Barreiras Channel, some 20 kilometres long, connecting the São Francisco River to the Marituba River and Lago dos Peixes. This natural channel is the route through which many species of fish reach the lakes within the marshland. The most important lake is Lago dos Peixes, known for its abundant fish resources. The vegetation of the wetlands is mainly marsh and includes several species of palm trees, used by the local population for building thatched houses, making medicine and obtaining food. The floodplain is the habitat for several species of fish, birds and small wild animals.

There are two villages in the floodplain – Marituba de Cima and Marituba do Peixe. Marituba de Cima has come into being recently as a result of expansion of the sugarcane plantations. Most of its 270 inhabitants have migrated from other regions, and work as wage-earners in nearby plantations. The people from Marituba do Peixe call it 'a place with no traditions'.

Marituba do Peixe, with a population of approximately 900, was established over a hundred years ago. The people's livelihood depends on traditional economic activities such as fishing, handicrafts and small-scale agriculture. The abundant fish caught in Lago dos Peixes are sold in Penedo, the largest nearby town.

During field work, over 48 different species of fish were identified, including cichlids, characins, catfish and several kinds of shrimp. In spite of the diminishing variety of habitats, several species of wild animals are still found – capybaras, agoutis and sloths, as well as caymans and snakes, and several species of aquatic birds.

The flora is mainly marsh, including varieties of pickerel-weed and sedge, some of which are used for handicrafts. Many species of trees are found on the islands in the lakes, including ouricuri palm, oil palm and *jatobá* (*hymenaea latifolia*). The trees provide wood for the construction of houses, fishing traps and canoes. Field work revealed over 30 different trees and shrubs used in the making of fishing equipment, and over 32 species used for medicine.

The *várzea* fishermen, drawing on a wide-ranging and precise knowledge of over 40 different habitats, have developed a series of management strategies and techniques to maintain the natural

146

productivity of the floodplain. The different lakes of the *várzea* are fished seasonally, but at different times. A resting period of four to six months is allowed for each lake, in order to enable stocks to recover, and this interruption of fishing is respected in the absence of any written laws or regulations. A second method of preserving fish stocks is the use of fish grouping techniques. One of the grouping techniques used to preserve fish stocks is called *manjuba*. Similar to the *akaja* technique used in West Africa, it consists of constructions of tree branches baited with cassava, which are laid in the water at specific points. Only adult fish are attracted, and thus the younger fish are preserved.

Far-reaching changes began to affect the floodplain and its people, the *varzeiros*, two decades ago. The first major impact was a change in the hydrological pattern of the floodplain as a result of the construction of two big hydroelectric dams hundreds of kilometres upriver. Before these dams were built, the São Francisco River flooded at regular periods. These floods, known locally as 'the waters of March', reached Lago dos Peixes through a natural channel, and brought with them several species of fish. When the flooding coincided with the ebbtide of the Atlantic Ocean, marine fish such as mullets, croakers and catfish were carried into Lago dos Peixes. The dams have regulated the flow of the river, so that there is now no difference in the water level between dry and flood periods. In the dry period the margins of the lake were formerly used for cattle rearing and subsistence agriculture, but with the constant higher water level these traditional activities can no longer take place.

As a result of these changes, the whole calendar of economic activities has been altered, deeply affecting the livelihood of the *varzeiros*. As they can no longer rely exclusively on fishing as a source of income, they must combine a wider range of economic activities, including handicrafts, coconut planting and, more particularly, wage-earning in nearby plantations. This combination of activities is now, in turn, being threatened by the expansion of the sugarcane plantations.

The second major impact resulted from the expansion of sugarcane plantations during the 1970s, as part of the government's programme for the production of alcohol to be used as car fuel. Although the region already had traditional sugarcane plantations, the Marituba floodplain was not used for alcohol production until the mid-1970s when, in response to government incentives, four alcohol distilleries were modernized and expanded their activities. As mentioned previously, Marituba de Cima is already surrounded by sugarcane plantations and the inhabitants are all wage-earners. Most of the land previously used for crops such as cassava and beans has now been taken up by sugarcane plantations. Cassava had traditionally been grown to make flour, as a main staple food.

In Marituba do Peixe, in spite of the people's resistance, the expansion of sugarcane brought deep changes to the floodplain, particularly in the land tenure system. The land here belonged to the Catholic Church and was leased to the peasants for a small rent. With the growing demand for land for sugarcane, the Church sold the land to Paisa, one of the distilleries, without informing the peasants and with no regard for their rights. With the assistance of the Rural Trade Union of Penedo, the peasants and fishermen were able to regain control over some small plots where their houses were, but lost most of their cultivable land. They do not even have the title to their housing plots, which still belong to Paisa.

The expansion of the sugarcane plantations has also had a deep ecological impact. Until recently, sugarcane was planted only on the high plateaux, but now it is being grown on the slopes facing the Marituba floodplain. The quality of the water is deteriorating as a result of erosion and siltation, and fish productivity is dropping. Intensive use of fertilizers and pesticides has also had a negative effect on fish stocks.

The last remaining areas of forest were cut to extend the sugarcane plantations. This deforestation has had a number of adverse effects on the local population. With the destruction of the habitats of various birds that were previously caught and eaten, the peasants and fishermen have lost an important source of protein. The disappearance of many fruit and palm trees, in particular the ouricuri palm, previously a source of straw for handicrafts, means that the women must now walk long distances to find such trees. A further consequence has been the increasing rarity of certain trees and plants that were used for medicine, and the difficulty in finding wood suitable for building the traditional fishing canoe.

The combined impact of the dams and the expansion of the sugarcane plantations, namely the absence of flooding and the constant higher water level, erosion, and pesticide and herbicide pollution, has seriously affected fish productivity. The ecological changes, in particular the increasing siltation and sedimentation, are undermining the fishermen's traditional knowledge of fishing areas and changing the landscape of the floodplain. Traditional management techniques, whereby some lakes were not fished from September to November when the water level was lower, are also less well respected. Lower fish productivity, and the fall in prices due to the disappearance of some of the larger fish, is leading to overfishing. The people of the wetlands are being forced to adapt their livelihood strategies in a way that takes them further from their traditional conservationist approach.

The latest and most serious threat to the *várzea* comes from CODEVASF, a government agricultural development agency which plans to transform the entire *várzea* into an irrigated area for intensive rice culture. This company has already converted several larger swamps of the São Francisco River into rice culture projects. The entire water regime of these areas is being changed. In the Betume project, covering 10,000 hectares, CODEVASF has closed off the water access to the lagoons and stopped fish migration. As a result, fish stocks have diminished and local fishermen have lost their livelihood. Apart from the serious environmental impact, the local population has also suffered from the conversion of the wetlands, having lost its land and been forced to live on the outskirts of the project area. The *varzeiros* were temporarily employed in the construction of the irrigated fields, but were seldom given a plot within the project area. Instead, the rice plots with irrigation infrastructure were given to better-off farmers and outsiders.

In 1985 CODEVASF decided to start a new project in the Marituba wetlands that would completely transform the *várzea*, disrupting the fisheries and the water regime. The peasants and fishermen would be resettled elsewhere. Marituba is now the last large floodplain not transformed by CODEVASF, and the population, with a long tradition of opposition to the latifundia, has strongly opposed this project. Various factors convinced it to oppose the CODEVASF plans. First, the example of several irrigation projects, such as Betume, in previously rich floodplains located close to Marituba showed that local populations gained very little from them. The *varzeiros* already plant small plots of rice, and they see no advantage in joining a large scheme, to which most of the farmers would become indebted. Second, the transformation of the *várzea* would destroy their traditional economy, which is based on fishing. Third, they would have to abandon their traditional houses to be transferred to a housing scheme where people from

149

outside would also be living. Their solidarity as a community would be disrupted.

Between 1986 and 1988 the people of Marituba do Peixe organized their opposition to the transformation of the *várzea*. They got strong support from the Fishermen's Association of Penedo and from a Catholic social service called Pastoral dos Pescadores. At the same time they were able to bring their problem to the attention of the Alagoas State Environmental Institute (IMA), which sent technical staff to the area and made several surveys, showing the ecological and cultural importance of the last floodplain of the São Francisco River. Finally, in March 1988, the state government declared Marituba an Area of Environmental Protection (APA). This environmental protection unit restricts large projects in the area; these can now only be implemented after an environmental impact assessment study.

The people of Marituba achieved their first goal, but the battle is not yet won. In early 1990 CODEVASF presented an environmental impact analysis to IMA for approval of the irrigation project. There is a risk that this study might be approved. The people of Marituba are concerned, because the present political situation is not favourable to them. They have started to get some additional support from external non-governmental environmental groups. A recent study carried out by the wetlands research and conservation programme from IUCN and the University of São Paulo, showing the great ecological and cultural importance of the *várzea*, is becoming an important tool in convincing the government and the IMA of the importance of conserving the Marituba floodplain and the culture of the local people (PPCAUB 1990).

CASE STUDY 2: THE RUBBER TAPPERS OF THE GUAPORÉ VALLEY

The Guaporé Valley is the largest area of wetlands, covering over 1,500,000 hectares, in the Amazonian state of Rondônia (see Figure 6.3), and is formed by the Guaporé River and its tributaries. The main wetlands area is located near the confluence of the Guaporé and Mamoré rivers, and has an average width of 60 kilometres, reaching 230 kilometres in the flooding period. The rainy season is from January to March, when the whole area becomes a large floodplain. The area is covered with dense forest and there are large areas of rubber trees and brazil nut trees. The main rubber tree

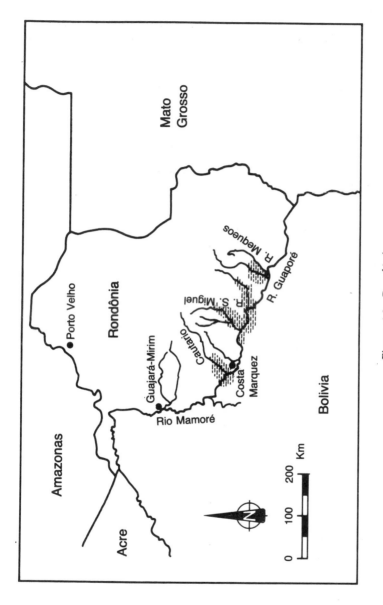

Figure 6.3 Rondônia

Source: *The International Atlas* (1991) Chicago: Rand McNally & Company

areas, or *seringais*, are found along the São Miguel River, where there are 20 *colocações*. These are socio-economic units of land worked by one or more rubber tapper families.

With approximately 100,000 inhabitants, the valley has a low population density of 0.64 inhabitants per square kilometre (1980). Half the population lives in the rural areas along the numerous rivers, and during the flood season some move to urban centres such as Costa Marques. There are sixteen Indian tribes living in the valley, with a total population of 3,179, most of whom live in indigenous Indian areas.

By the end of the last century, a railway had been constructed linking the south of Rondônia to the capital, Porto Velho. At that time, most of the rubber extraction was done by the Guaporé Rubber Company, using British capital. The construction of the railway attracted workers from the north east and also ex-slaves, many of whom died from malaria and other tropical diseases.

One of the region's main economic activities is fishing, particularly in the Guaporé, Cautário and São Miguel rivers. Until recently most of the fishing was done on a subsistence basis, but now large commercial fishing boats are coming to the area from Porto Velho. A greater threat is pollution from the mercury used in gold mining in the Mamoré River, which might also reach the Guaporé Valley. Other essential activities are rubber tapping and harvesting brazil nuts, which are done by both the Indians and the riverine population.

Extraction activities are still the backbone of the region's economy, generating more income than cattle rearing, which is heavily subsidized by the federal government. The Brazilian term *extrativismo* refers to the removal of non-timber forest products such as latex, resins and nuts, without felling the trees. Approximately thirty products are collected, on a sustainable basis, for commercial sale, while many other forest materials are gathered by the 'extractivists' for their own use as food and medicine (Fearnside 1989).

Rondônia still has a high proportion of independent rubber tappers, and they have one of the strongest extractivist organizations. Most rubber tappers elsewhere are still under the '*aviamento*', or debt peonage system. Under this system they sell their products and buy their provisions from a rubber baron, or *seringalista*, contracting ever-increasing debts which then hold them captive.

As a result of the recent occupation of the Guaporé Valley by farmers and logging industries from the south, the indigenous population is being pushed out, in particular by the large latifundia that use fiscal incentives for establishing cattle ranches. Many rubber tappers from the area have crossed the border to live and work in Bolivia.

In the neighbouring state of Acre, over 500,000 people earn their living collecting latex from wild rubber trees. At the beginning of the 1970s, the *seringueiros* here organized themselves into unions. They started peacefully resisting eviction from the *seringais*, 'sitting down' in front of the timber saws to prevent the trees from being felled. The National Council of Rubber Tappers, created in 1985 for the conservation of the Amazon forest, has met with a great deal of opposition from the strong economic lobbies of powerful land-owners. Many leaders, including the well-known Chico Mendes, were brutally killed. The organizational work of the National Council has recently reached the Guaporé Valley, and the first rubber tappers' meetings were held in Guajará-Mirim in 1988.

The death of Chico Mendes brought massive national and international support for the struggle of the *seringueiros* against the destruction of the Amazon forest. It also made clear that the tropical forest cannot be protected and used on a sustainable basis if the interests of traditional communities are not taken into account, particularly those of rubber tappers and Indians. In this context, the strategy chosen by the National Council of Rubber Tappers, based on the creation of 'extractive reserves' is very important.

The concept of the extractive reserve is much more than an ecological issue. The *seringueiros* are fighting for the forest, not necessarily because of the forest itself, but because it is the source of their livelihood and their very way of life. Without the forest there would be no latex, brazil nuts or other forest products on which they depend for survival. The establishment of these reserves requires true agrarian reform in the Amazonian region, which is presently obstructed by the conservative forces in Brazil.

According to the proposed system, the extractive reserves are to be communally run, with the government retaining land ownership in a manner similar to Amerindian reserves or parks. This is not intended to be a form of resource collectivization. Although not issued separate deeds, individual families retain their rights to tap in their traditional *colocações* within the reserves. The land cannot be sold or converted to non-forest uses, although small clearings for subsistence crops are permitted, usually not exceeding 5 hectares per family or approximately 1 to 2 per cent of a reserve (Fearnside 1989).

Each *colocação* might have more than one track, covering an average of 500 hectares. The *colocação* is not only a number of tracks linking rubber trees tapped by one *seringueiro*, but a productive unit in which the family does subsistence farming and hunting. In some instances, by reducing the size of the *colocação*, the *seringueiros* have been forced to sell their land and go to towns as wage-earners. This is why the National Council of Rubber Tappers is fighting for extractive reserves to include a number of *colocações*, and not to consist only of small individual plots.

The extractive reserve proposal should not be viewed as a means of supporting a dense population or of absorbing people migrating from other regions. Those new to 'extractivism' lack the knowledge and attitudes necessary to make the system work on a sustainable basis. Even with experienced rubber tappers, only a sparse population can be supported. This is currently estimated at 1 to 1.7 persons per square kilometre. Great care must be taken that the terms 'extractive project' or 'extractive reserve' do not become mere euphemisms for the type of settlement that has already been discredited on the Transamazon highway and in Rondônia, thus indirectly discrediting legitimate extractivism (Fearnside 1989).

Economic self-sufficiency is an important goal for the extractivists. It will require maximizing the variety and value of products sold, limiting the loss of money to intermediaries and minimizing the cost of establishing and maintaining the reserves (Fearnside 1989).

Another important element in the establishment of extractive reserves is their economic viability. It is known that the latex produced in natural forests is more costly than that produced by artificial rubber plantations. However, as Alegretti (1987) has shown, costs are calculated according to standards set by the traditional rubber enterprise. According to this *aviamento* system, the *seringueiros* are totally dependent on the food brought from outside by the *seringalistas* and sold to them at very high prices. The independent rubber tappers, although paying 10 per cent of their income to the landowner, produce their own food in the *colocações*. Production costs are therefore lower for an independent rubber tapper than for the tappers trapped in the *aviamento* system.

If the *seringueiros* can manage to establish their own cooperative societies and small latex processing units, they will be able to solve their main problem, which is the low selling price of their products. Marketing mechanisms for new products also need to be developed if extractivists are ever to enjoy a reasonable standard of living. As the poverty of the rubber tappers during the rubber boom has shown, when the real value of the products is appropriated by intermediaries, extractivists remain poor, regardless of the amount of wealth they have generated.

The rubber tappers are now making a major effort to diversify the products they extract and sell. This strategy would require collaboration with researchers, such as pharmacologists, chemists and botanists, who could develop new products, especially from medicinal plants. By limiting the products collected and the extent of harvesting, care must be taken that only sustainable extraction is practised (Fearnside 1989). In order to diversify the economic activities within the extractive reserves, the National Council of Rubber Tappers has recently created a training and research centre. They have recruited forestry engineers who are studying ways of increasing the number of rubber trees in the colocações.

Another issue to be addressed is the legal aspect of the establishment of extractive reserves. There is no existing model in Brazil for the establishment and management of a protected area, which meets the objectives of the sustainable use of natural resources. All existing categories require that no one live in an environmental protection area. Even the so-called 'national forest', where logging is allowed, is not suitable as a model because this was created to respond to the interests of large logging enterprises. When a national park or ecological station is created, the local population is moved out (Alegretti 1987).

There are several examples of national parks or biological reserves being established in places where traditional communities of fishermen and small-scale farmers previously lived. These communities, usually living in isolated areas, depend exclusively on natural resources for fishing, fruit harvesting, subsistence hunting and other livelihood needs. They establish a complex relationship with the natural environment, which is not only of an economic nature. Values, traditions and cultural perceptions established over centuries play an essential role in defining their relationship with the environment and its natural resources. Very often the government declares an area protected without taking into consideration the importance of its role as a local heritage. The establishment of strict protection for large areas has led local communities to social revolt as their means of livelihood is abruptly suppressed. In consequence, they consider the new protected area as common land and start to over-use natural resources and hunt game illegally, which they did not previously do (Diegues 1989).

Without the traditional users, the areas suffer severe degradation from logging and mining companies. The presence of traditional communities can therefore be considered as an assurance of continued conservation, provided that an adequate management scheme is found and accepted by local communities. Traditional schemes can also be a source for the formulation of improved management techniques. In view of these problems, the *seringueiros* are proposing a new type of protection area where the sustainable use of natural resources is permitted. At present, 14 extractive reserves have been approved: 7 in Acre, 4 in Amapá, 2 in Amazonas and 1 in Rondônia. The reserves cover three million hectares and have a population of 9,000 families.

CONCLUSIONS

1 In both case studies, traditional communities have an economic structure based on the intensive but sustainable use of natural resources.

2 The capitalist mode of production is penetrating both communities. In Marituba, the expansion of sugarcane plantations on the *varzeiros'* land is forcing them to become wage-earners. In the Guaporé Valley, the expansion of cattle ranching and logging is destroying the forest from which the *seringueiros* make their living. This process is forcing them to move to the slums of the towns.

3 The impact of the penetration of the capitalist mode of production is devastating the ecological system of the Guaporé Valley and the Marituba floodplain. The hydrological regime of the Marituba floodplain has been changed by the construction of hydroelectric dams and the establishment of modern irrigation. Fish production, on which the fishermen and peasants depended for their living, has been decreasing, reducing the meagre income of the *varzeiros*. In Rondônia, the expansion of cattle ranching is dramatically reducing the size of the forest, eliminating the rich biological diversity, and the different habitats of animals and birds. The deforestation is also destroying the basis of the important local extraction activities based on rubber, nuts and pharmacological products.

4 In both cases, local populations have strongly opposed the

destruction of their livelihood. It has become clear that the protection of their livelihood is a prerequisite for the conservation of the natural environment. It is from this perspective that both communities are actually proposing, through similar strategies, the continuation of a sustainable use of natural resources, and not simply the preservation of the forest as a national park.

5 Conservation practices are embedded in the structure of both traditional communities. These practices spring from a deep knowledge of the natural environment and its processes. The socio-economic changes described earlier are disrupting these practices.

6 Both communities have proposed the establishment of environmental protection units in order to stop the destruction of the natural environment of the floodplain. In the case of Marituba, the environmental unit established by the government does not necessarily prohibit the transformation of the floodplain. However, all major development projects require an environmental impact analysis. In the case of the Guaporé Valley, where extractive reserves are proposed, the legal situation of this unit has not yet been resolved.

7 In both cases, it has become clear that, without strong social mobilization, neither the livelihood of the traditional communities nor the environment could have been conserved. In the case of Marituba do Peixe, the mobilization was only locally based, but the inhabitants succeeded in gaining support from the regional Fishermen's Union. The rubber tappers of the Guaporé Valley were able to gain wider national and international support for their struggle. Many *seringueiros* leaders, including Chico Mendes, were murdered by landlords who see the extractive reserves as the beginning of agrarian reform, and of sustainable development in Amazonia.

NOTE

1 Report based on comparative field studies conducted by interdisciplinary teams under the Programme on Research and Management of Wetlands in Brazil (University of São Paulo/International Union for Conservation of Nature and Natural Resources/Ford Foundation).

REFERENCES

Alegretti, M. (1987) *Reservas Extrativista; uma proposta de desenvolvimento da Floresta Amazônica*, Curitiba: Instituto de Estudos Avançados, mimeo.

Bourgoignie, G. (1972) *Perspectives en Ecologie Humaine*, Paris: Editions Universitaires.

Diegues, A. (1983) *Pescadores, Camponeses e Trabalhadores do Mar*, São Paulo: Editora Atica.

Diegues, A. (1989) *Application of the Biosphere Reserve Concept to Coastal Marine Areas in Brazil*, São Paulo: University of São Paulo.

Fearnside, P. (1989) 'Extractive reserves in Brazilian Amazonia', *BioScience* 39 (6).

Forman, S. (1970) *The Raft Fishermen: tradition and change in the Brazilian peasant economy*, Indiana: Indiana University Press.

PPCAUB (Programme on Research and Management of Wetlands in Brazil) (1990) 'As várzeas ameaçadas: um estudo preliminar das relações entre as comunidades humanas e os recursos naturais da Várzea de Marituba-Rio S. Francisco', São Paulo: University of São Paulo.

Part III

SOCIAL ACTION AND THE ENVIRONMENT

7

URBAN SOCIAL ORGANIZATION AND ECOLOGICAL STRUGGLE IN DURANGO, MEXICO

Julio Moguel and Enrique Velázquez[1]

This chapter discusses the origins, goals and strategies of the Comité de Defensa Popular (CDP) of Durango, Mexico, with particular emphasis on the development and accomplishments of the environmental programme of activity undertaken by this popular social movement. It opens with a discussion of the context within which the predecessors of the CDP in the Durango area were formed, and goes on to analyse the process by which members of the organization became convinced that attention to ecological issues was essential to the social, economic and physical health of the community, and to discuss how the organization was able to make an impact in this area. It is argued that the popular movement which was eventually successful in addressing many of the environmental problems of the region originated with a concern for the social and physical well-being of the community, and retained its focus on its social point of departure: it has never centred its activities around a purely 'environmentalist' concept as such.

HISTORICAL CONTEXT

Northern Mexico, which includes the state of Durango, shares a political, social and economic background which distinguishes it from other areas of the country. From the north emerged the principal social, political and military factions of the Mexican revolution of 1910–17, and it is largely to these elements that modern Mexico owes its existence. It was in the north that the first significant popular struggles against the Porfirian regime took place,

161

such as the Cananea strike and the attempted insurrections of the *magonismo* movement of radical proponents of agrarian reform. It was also in this region that the dominant ruling bloc at the end of last century suffered its first significant schisms (Madero, Carranza and Sonorenses), opening the way for the popular uprisings which together with constitutionalism and *villismo* (the following of Francisco Villa) led to the fall of the dictatorship.

In modern times the first important popular struggles of the north took place in 1957–8, with the simultaneous invasion of lands in the states of Sonora, Chihuahua and Sinaloa, led by the Unión General de Obreros y Campesinos de México (UGOCM: General Union of Mexican Workers and Peasants). More than their effectiveness and success, the significance of these mobilizations was linked to their taking place outside the framework of, and counter to, the traditional corporative structures built up by the revolutionary regimes since the 1930s. It was at the same time that the Partido Revolucionario Institucional (PRI: Institutional Revolutionary Party)[2] suffered its first great electoral setback in a federal state, when the Partido Acción Nacional (PAN: National Action Party)[3] won the elections in Baja California. This victory was not recognized by the government, causing a social conflict of unexpected dimensions, and culminating in army intervention and a massive and bloody repression.

Between 1960 and 1964, further important popular struggles developed, basically agrarian or rural in nature, all with the common feature of deriving from social sectors which had broken away from government-controlled organizations, in particular the Confederación Nacional Campesina (CNC: National Confederation of Rural Workers). These northern regional movements then joined the organizational plan of the Movimiento de Liberación Nacional (MLN: National Liberation Movement), whose constituent groups became its fundamental source of social cohesion.[4] In 1963 many of these agrarian groups joined to form the Central Campesina Independiente (CCI: Independent Rural Workers' Headquarters), the first rural group of national importance to be created outside the framework of, and counter to, the traditional corporative organizations, in particular the CNC.

During the first half of the 1960s other events shook the north of the country. A guerrilla group emerged in Chihuahua in 1963 with the aim of confronting the problem of enormous and growing social inequalities, and in particular of fighting against the old *cacicazgos*, or *cacique* systems, in the region (*caciques* were originally Indian chiefs, and by extension any powerful non-institutionalized local leader who ruled the political and economic life of the community). On 23 September 1965, this armed group attacked military barracks in Ciudad Madero, and as a result was practically wiped out.

At this stage further political events assumed importance. In 1967, PAN took advantage of a sharp division between the dominating classes and sectors in Sonora to deal a spectacular electoral blow to the PRI by winning control of the municipalities of Hermosillo, Bacoachi, San Miguel Horcasitas, Cucurpe, San Pedro de la Cueva, Santa Ana and Opedepe. In the wake of government recognition, these triumphs constituted the first significant electoral victories for the opposition, thus foreshadowing a pattern of contradictions which gradually developed throughout the 1970s and 1980s within a much wider regional framework.

At the beginning of the 1970s, the economy and the politics of the north of the country began to undergo far-reaching changes. The flourishing assembly plant industry and the setting up of a new automobile industry (with a high degree of modernization to meet the needs of the North American market) strengthened the trend towards the integration of Mexico with the United States. Large and medium-sized towns of the region experienced a new and vigorous growth cycle, caused as much by the new industrial development trends as by migration from rural to urban areas, a product of the economic crisis in the countryside.

Towards the end of the 1970s a series of factors provoked the creation of new civic groups and social movements in the northern states of Mexico. Of particular relevance were the events taking place in Durango in 1966 and 1970 in two separate cycles of mobilizations, involving a multisectoral and poly-classist campaign against exploitation of the state's mineral resources by the industrial magnates of Monterrey, who bought crude iron from Cerro del Mercado at an extremely low price, thus depriving the people of Durango of the potential means to start up their own industrial structure. The national student mobilization of 1968 and its tragic consequences generated fresh disquiet in the popular sectors of the north, radicalized many of its ideological positions and provoked further splits away from the traditional structures of corporative power.

It is within this context that new popular groupings emerged: in

Chihuahua, the confluence of formerly independent syndicalists, peasants fighting for land and productive reorganization, and students and *posesionarios* (squatters, or 'invaders' of land) led to the foundation at the beginning of the 1970s of the Comité de Defensa Popular de Chihuahua (Chihuahua Popular Defence Committee), which years later came to be made up exclusively of *posesionarios*. In the north east, a process of successive invasions of urban lots by migrants arriving to the city from the countryside led to the formation of the Frente Popular Tierra y Libertad de Monterrey (FPTyL: Land and Liberty Popular Front of Monterrey). These were the first two important pressure groups organized to work on behalf of *colonias*[5] to be formed in the country.

Between 1975 and 1976, a series of nearly simultaneous land invasions by peasants from Sonora and Sinaloa provoked one of the most acute social and political conflicts affecting the modern history of the country, leading to the fall of a governor and a profound split between wide sectors of the northern land-owning and industrial middle class and the ruling political bureaucracy. The then President Luis Echeverría Alvarez (1970–6) decided to expropriate hundreds of hectares of irrigated land to satisfy the demands of some of the peasants. Governor José López Portillo (1977–82) compensated the expropriated owners and anxiously sought to rally the dominant bloc together again. Despite his efforts, however, an important part of the northern agricultural and industrial middle class remained politically dissident. This was reflected in the new electoral trend: from now onwards PAN was able to capitalize on some of the economic and political favours of important centres of economic power and of the church, a trend which was re-established and extended in 1982, when Mexico entered into a long and difficult cycle of economic crisis, and when the President unpredictably and without the prior agreement of the ruling sectors decided to nationalize the Mexican banking system.

In the federal elections of 1982 the PRI barely achieved 68 per cent of the votes at the presidential level, representing the party's lowest ballot in electoral history since the formation of the Partido Nacional Revolucionario (National Revolutionary Party, a forebear of the PRI) in 1929. Though to a lesser extent than in other states, this loss of support for the PRI was also reflected in the elections in Durango, where its share of the votes dropped from 83.4 per cent in the 1979 elections to 74.9 per cent in 1982. At the national as well as the local level PAN was the principal beneficiary of this swing in electoral preference: at the national level PAN went from 10.8 per cent of the votes in 1979 to 17.7 per cent in 1982; the figures for Durango for the same years being 8.4 per cent and 18.3 per cent respectively.

This new electoral trend finally precipitated in the local elections which took place in July 1983: in Chihuahua, PAN triumphed in 7 of the principal municipalities, and in Durango, with 38,931 votes in its favour against the 30,016 of the PRI, it gained control of the government of the state capital. PAN also carried off 2 of the 13 seats for deputy. The federal elections of 1985 confirmed the basic direction of the trend foreshadowed in 1983: in Durango, the PRI saw their percentage of the votes drop to 66.4 per cent, with PAN obtaining 26.2 per cent, their highest gain in state elections. The significance of this figure can be more clearly appreciated with the added knowledge that it was the second highest result for PAN in the whole republic, after they had achieved only ninth place in the 1982 elections.

THE ORIGINS AND DEVELOPMENT OF THE DURANGO POPULAR DEFENCE COMMITTEE

Almost simultaneously with the formation of organizations such as the CDP in Chihuahua and the FPTyL in Monterrey, the seeds of what was later to become the Comité de Defensa Popular (Popular Defence Committee) began to emerge in Durango. The recent antecedents of this organization were the above-mentioned mobilizations of the Cerro del Mercado in 1966 and 1970, and, at the national level, the student movement of 1968. Its 'structural' background lay in the fact that the economy of Durango was based fundamentally on agriculture, livestock and forestry, with an important role played by the sharecropper sector. The state's industrialized sector never managed to take off,[6] and Durango continued throughout the 1970s to have one of the highest rates of rural–urban migration in the country, primarily because of the crisis and the decreasing rhythm of capitalization in rural areas (Zacatecas was the other state affected by such high migration rates). The principal towns (Durango, Gómez Palacio and Lerdo, see Figure 7.1) grew mainly on the basis of the services sector, offering only limited employment opportunities for the expanding population. Added to this, the close of the golden era of the Mexican economy was beginning to affect those urban centres which were already

165

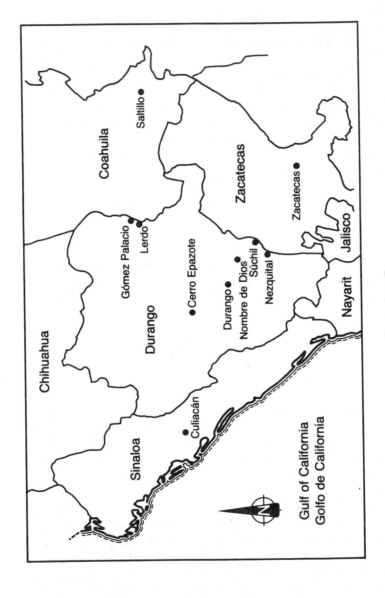

Figure 7.1 Durango

Source: *The International Atlas* (1991) Chicago: Rand McNally & Company

established but which lacked resources, and where protests now began about the total lack of, or the increasing cost of, public services.

The first organizational nucleus of the Movimiento Urbano Popular (Popular Urban Movement) in Durango began functioning in the city of Durango in 1972. It consisted of activists from Mexico City who had participated in the student movement of 1968, and who formed part of a political and ideological current self-defined as maoist, which proclaimed the need for revolutionary socialist change. Such change was to be brought about by the 'broad popular masses' through a prolonged process (following the idea of the prolonged popular war of maoism), which would advance from 'the countryside to the city', and from 'the periphery to the centre'.[7] After some minor failures in their efforts in the countryside they moved to the capital of the state, where they made connections with tenant groups fighting to reduce the water rates. Their first triumph converted them into a group of popular composition which aimed to undertake an initial invasion of urban lots to found a *colonia*. This action, which took place in an old airfield, was brutally suppressed. A similar fate befell a further invasion, planned for 9 March 1973. The repression soon became selective and led to the temporary detention of several leaders. However, it was not long before the group managed to establish a territorial 'base'; in the second half of 1973, political negotiations with the federal dependency then in charge of housing (INDECO) enabled the activists to acquire lots on which the first *colonia* of the group was founded, called División del Norte (Northern Division).

After the establishment of the first popular *colonia*, the work of the activists (or of the already organized popular nucleus) developed along two basic lines: that of building up internal democracy within the *colonia* and establishing new *colonias* in order to meet the increasing demands for land and housing, and that of broadening the sphere of action and political force of the 'movement'. As regards the first, assemblies were created for each sector and the first internal organs of local governing structures were designated. The General Assembly of the *colonia* was set up as the highest authority. In following the second line, three years after founding the first territorial base a new cycle of invasions was initiated, and as a result of the first two successes the *colonias* Emiliano Zapata and Lucio Cabañas were formed.

During the years 1976–9, the situation among the rank and file

167

of the three newly founded *colonias* was complex. In División del Norte, internal divisions appeared, and a series of conflicts led to the fracturing of the *colonia*'s organization. In the other two *colonias*, problems developed in relation to the lack of consolidation of the organized nuclei or sectors. A process of struggle, which was to a certain extent 'external' to the organized nuclei in the three *colonias*, led to a qualitative change in their development: a popular mobilization against high electricity tariffs brought together groups from about twenty *colonias*. It was this development which precipitated the process of reorganization and lent it dimensions that nobody at the time could have imagined: a popular assembly of the groups involved decided on the formation of the Comité de Defensa Popular (CDP) General Francisco Villa (General Francisco Villa Popular Defence Committee).

Between 1980 and 1985, the CDP expanded and became consolidated, assuming the two lines of work contemplated initially: that of building up internal democracy and that of territorial and sectoral advance or extension (in the form of self-extension or supporting other groups and popular sectors, with the idea of 'using internal lines of work to back up external lines', and vice versa). Gradually, the following *colonias* were formed by either the invasion or the purchase of lots: Tierra y Libertad, Genaro Vázquez, Arturo Gámiz, José Revueltas, 10 de Mayo, Manuel Buendía, Isabel Almanza, 8 de Septiembre, Jalisco and Hipódromo.

The 'ideological struggle' began to take on certain cultural dimensions: apart from community life and work (such as the activities organized for 'red Sundays', the general assemblies and group work), certain rules of coexistence were laid down whereby, for example, alcohol gradually disappeared from the big popular celebrations. The effect of these rules was also felt in the spheres of basic education and health: medical clinics were set up, and study circles and supply centres were created. Auditoriums were constructed or established, and popular communication programmes were developed. In 1983 the Centro Cultural José Revueltas (José Revueltas Cultural Centre) was founded in an attempt to formalize and centralize other activities in this same vein. This period saw the CDP pressing ahead in the development and consolidation of its organizational structures. In 1982 the first Comisión Política General (General Political Board) was designated, and rank and file political boards were later set up at the *colonia*, sector and committee levels. At the same time, and without ceasing to be an organization

168

composed fundamentally of *colonias*, the CDP managed to integrate core groups from other social sectors such as small-scale merchant affiliations, and even syndicates such as that of musicians, which had split away from the corporative structure of the CTM. The CDP became active in the other two main towns of the state (Gómez Palacio and Lerdo) and began working with peasant groups from Coneto de Comonfort, Súchil and Nombre de Dios. It was during these same years of promotion and consolidation of the CDP that the Coordinadora Nacional del Movimiento Urbano Popular (Conamup: National Coordinating Committee of the Urban Popular Movement) was set up, in which the CDP of. Durango played a fundamental and key role.[8]

The year 1985 was nevertheless one of serious difficulties for the CDP; as it began to regard this new social organism as both a real and potential enemy which was expanding and strengthening in geometrical progression, the state government embarked on a new cycle of repression. In the same year an armed provocation was mounted in the José Revueltas *colonia*, resulting in the assassination of José Angel Leal and the shooting of six CDP members. At the beginning of 1986 the secretary general of the musicians' syndicate, Juan Lira Bracho, was assassinated.

In the face of this situation the CDP proposed moving on to a new phase of development, under the motto of 'breaking the siege', and progressing to other planes of political activity. This implied abandoning a hitherto central aspect of CDP policy, namely a reluctance to participate in the electoral process, considered the *lebensraum* of the PRI-bourgeoisie within the political arena. The idea prevalent in the past was that popular movements ought not to enter into this terrain 'on their long road of revolutionary struggle'.[9] Consequent to this change of tactics, the CDP participated in the 1986 elections:

> in order to build up new defences in the face of political repression by the government, open up channels of communication with other sectors of civilian society in Durango, and remodel the struggle for civil rights and self-government at the local or *colonia* level with a view to participating in the management of certain areas of public administration, above all in the municipal sphere.   (Moguel 1990)

But there were other reasons too: the increased strength of PAN in the north of the country generated incompatibilities with the PRI, a situation which could be used to advantage by the CDP with the promise of rapid and significant dividends. PAN had managed to seize control of the most important municipality of the entity from the PRI (the state capital, from 1983 to 1986), and there was everything to indicate 'rough waters' ('río revuelto'[10]) in the next elections.

The results were surprising, taking into account the traditional weakness of the left in electoral processes in the north of the country and the tradition of non-participation of CDP supporters: in the 1986 elections, with around eight thousand votes, the CDP captured a deputy seat in the local congress and several *regidurías*.[11] Thus began a qualitatively new phase in the life of the CDP. Their first 'parliamentary' experience in municipal administration acquired particular relevance and came as a surprise to the 'political society' of Durango: they showed a rare capacity for coming up with proposals, they were constantly on the initiative, and they made an impact with their novel methods and styles of action and work. Some of the meetings of the local Congress were 'filled with CDP supporters' backing up the participation of their representatives, or putting pressure on the PRI and PAN speakers. In the case of the municipal government of Durango, the CDP *regidor* took his administrative tasks very seriously and in a very short time his offices were converted into an obligatory meeting place for CDP supporters and for citizens in general who knew that they would always be listened to, and that many of their problems would be resolved. The local press began to record, on daily basis, the proposals and political and social activities of the CDP, whose members gradually provided a compulsory reference point in discussions on what was happening in Durango and on ways of solving the various problems in hand.

With these changes, the CDP itself underwent important transformations: in its internal life, traditionally organized on the basis of direct democracy through assemblies, the secret ballot and universal suffrage were introduced for the election of all the leadership posts. At the same time a framework was set up to provide technical support in various areas of work. The Guadalupe Victoria Legal Office was set up to consolidate and formalize the activity of lawyers who until now had only given circumstantial or discretionary support.

From 1988 to 1989 the CDP continued to participate in the electoral process, winning a federal deputy seat (with its top leader as deputy), two local deputy seats, and *regidurías* in five municipalities within the state. The party also won municipal elections in Nombre de Dios and Súchil as a culmination to several years of activity in rural areas close to the town of Durango, and above all as a result of the commencement of mass activity in the ecological sphere against the contamination of the River Tunal (discussed further below).

The ultimate milestone marking the culmination of the above developments was the signing of an agreement between the CDP and the federal government in March 1989. With this agreement the CDP had succeeded in winning some of its more pressing demands, such as the installation of popular kitchens, a building materials bank, 2 productive workshops, 5 tortilla plants (in the towns of Durango, Gómez Palacio and Lerdo), a centre for child development, the renovation of classrooms, and the installation of electricity and paving in some *colonias*. The CDP now faced the challenge of proceeding to a new and complex phase of social and productive self-organization, surpassing the boundaries of the *colonia* – or of the sum of *colonias* and grassroots groups – to take on the administration and control of a complex network of productive relations.

This new situation did not mean that efforts to extend the territorial base of the CDP were abandoned. In addition to *colonias* such as José Angel Leal and Juan Lira, plots of land were 'invaded' or 'negotiated', permitting the creation of the Hipódromo and Canelas (this name was later changed to José Martí) *colonias*. By the beginning of 1990 the CDP had more than 60 grassroots groups distributed throughout the towns of Durango, Lerdo and Gómez.

In the meantime, the economic difficulties of the rural areas of Durango and the contamination of the River Tunal had enabled the CDP to extend its work of social promotion and mobilization to the peasant sector, so that they were not only able to win control of the municipalities of Nombre de Dios and Súchil, but also to form a specific organization of communities and *ejidos* called the Unión de Pueblos Emiliano Zapata (UPEZ: Emiliano Zapata Union of Rural Settlements).

THE GROWTH OF THE ENVIRONMENTAL MOVEMENT

The phenomenon of contamination in the River Tunal and the concomitant environmental deterioration within the municipalities and other areas through which the river passes date back to the second half of the 1970s, when the Alfa group (one of the most powerful industrial groups in Mexico) installed the Celulósicos Centauro enterprise (a cellulose plant) on the banks of the river. The production of cellulose and its derivatives is known to be a highly contaminating process because of the type of chemical substances used. It was not only the damaging effects inherent in the production of cellulose which gave rise to the grave contamination, but the additional fact that the industry operated on imported second-hand technology which did not include the necessary installations for the treatment of chemical waste.

The social protest against contamination did not begin immediately; it appeared some time after the pollution began, originally in a rather spontaneous and disjointed form because of the initial difficulty in detecting the basic causes of the pollution and the prevailing inexperience in confronting this type of problem on the one side, and governmental indifference and the economic and political influence exerted by the owners of the enterprise on the other.

Contamination and its repercussions worsened at the beginning of the 1980s, when, in addition to the residual waters from the cellulose industry, sewage from the town of Durango began emptying into the bed of the River Tunal at a rate of 1,000 litres per second, and without adequate prior treatment. The reason for this was that the existing oxidation pools were by now inadequate and obsolete, affected by the mere passing of time, the lack of investment and technical maintenance, and the quite rapid population growth in the area. Little by little the problem became evident to the population at large: the bottlenecks along the drainage networks multiplied, and 'El Saltito', part of a small waterfall on the River Tunal which had once been a favourite recreational spot for local inhabitants, was gradually abandoned.

It was the CDP which began to take the problem seriously. Building on their solid standing in the town and their links with some popular groups in the surrounding rural areas, they began to mobilize organized protest. In July 1987 they called for a mobilization, taking advantage of the juncture of two events: the cellulose factory changed hands and was now owned by the Durango Industrial Group, and the then head of the Secretaría de Desarrollo Urbano y Ecología (SEDUE: Secretariat for Urban Development and Ecology), Manuel Camacho Solís, paid a visit to the state. The mobilization, bringing together several hundred peasants from the region, and consisting in the blockage of the Durango–Mexico City motorway, had only one demand, but a radical one: the closure of the plant causing the contamination. Under instructions from the head of SEDUE to deal personally with the conflict, one of the sub-secretaries paid a visit to the plant and to the river accompanied by a large group of *ejidatarios*, and was able to confirm the magnitude and gravity of the problem. The result of the visit was a formal undertaking on the part of the owners of the factory to install the necessary technology for the adequate treatment of chemical waste within a period of two years.

This mobilization and incipient series of activities related to the problem of contamination achieved other important objectives, in particular that of eliciting a compromise on the part of the head of SEDUE to start on the construction of a plant to treat the contaminated waters of the city of Durango. Within this context there developed an initiative aimed at forming an organization to promote action and coordinate the various political and social sectors concerned with problems relating to the conservation of the environment. As a result of this initiative, the Comité Duranguense de Defensa y Preservación Ecológica (Durango Committee for Ecological Defence and Preservation) was set up, bringing together non-party citizens, the CDP, business community groups (of the Canacintra), and sectors attached to various political parties (including PRI and PAN, and militants from the Partido Mexicano Socialista (Mexican Socialist Party).

The formation of the Comité Duranguense de Defensa y Preservación Ecológica was without precedent in the state: never before had contamination been seen from an ecological perspective, and still less had there been any previous attempt to coordinate multisectoral action of a similar magnitude and impetus with a view to resolving comparable problems. What most surprised many people was that the initiative had come from the CDP (some sectors still regarded this body as being behind the times and incapable of showing initiative or coming up with solutions to general problems), and furthermore, that it was accepted and adopted by the political parties and business community.

But the Comité Duranguense was very short lived, not only because of the difficulties involved in coordinating and maintaining unity among such disparate sectors, but also because neither the project itself, nor the ideas and proposals sustaining it, matured properly: the two years' grace given to the cellulose plant owners to install anti-contaminatory equipment, and the promise of government officials to undertake medium-term measures to solve the city's sewage problems (the date set for complying was 1989), contributed to make the organizational process lethargic and to detract from its consistency. In the course of a few months the fabric had unravelled and few had the heart and energy to press the matter or to call for new mobilizations. By 1988 little was left of the Durango Committee, and later attempts at organization started afresh from a new base and conditions.

Unlike the rest of the groups involved, the CDP insisted on continuing the process; the problem had not been adequately dealt with by the mere extraction of promises from businessmen and government officials, and they were convinced of the need to take the initiative in developing 'an ecological conscience among citizens' and seeking 'intermediate solutions'. Thus in 1988 they set to work reviving the project, promoting a diagnosis of the levels of contamination in the River Tunal as well as its effects on production and on the health of the inhabitants of the affected areas. By this time the Unión de Pueblos Emiliano Zapata (UPEZ: Emiliano Zapata Union of Rural Settlements), had already come into existence, a sister organization to the CDP comprising various peasant groups from rural areas in the neighbourhood of Durango City, which wholeheartedly took on the struggle 'for ecology'.

This endeavour led to the formation of a new organizational structure, made up initially of nuclei of rural workers from the affected zones, a group of technicians and activists expert in ecological issues, and some ten non-party citizens interested in supporting activities and mobilizations of this nature. The new organism was called Comité de Defensa y Preservación Ecológica (CDyPE: Defence and Ecological Preservation Committee), and it retained the idea of the previous Durango Committee of developing as a pluralistic organization independent of other social and political bodies, including the CDP.

The experience accumulated during the lifetime of the original Durango Committee convinced the new one that they should adopt a different approach. The first thing they did was to register the CDyPE as a legal entity. The task of diagnosing the extent of contamination in the River Tunal and its effects on production and health in the affected zones was continued, and a process of consultation and discussion was initiated to stimulate the rapid implementation of alternative 'intermediate' measures, and to reverse at least partially some of the graver effects of contamination. Another immediate objective was to acquire premises where the committee could convene and function without having to depend politically on the CDP or any other body which would detract from its independent standing and ability to grow in a diversified and multisectoral manner.

The specific objectives put forward were to: 'organize the citizens in a manner conducive to their active participation in the solution of ecological problems in the rural and urban environment of the State of Durango'; devise alternative proposals for social and productive development; and 'assist particularly in resolving the problem of contamination in the River Tunal, which is prejudicial to the producers and inhabitants' of various municipalities. The chosen strategies were 'the permanent undertaking of technical-scientific research to provide full knowledge of the facts to the citizens in devising their ecological policy'; to inform and train those sectors interested in the ecological problems of the state; to design and stimulate structural and intermediate solutions to the various problems connected with the environment; to create CDyPEs 'wherever necessary and feasible'; and 'to build up and foster a professional team of scientists and technicians specialized in environmental issues'. The committee was soon able to establish its own dialogue with the government and with the business community in Durango, and a process of multilateral negotiation for obtaining economic resources was initiated. A radio and press 'campaign for ecology' was begun, using the slogan 'a healthy environment is a fundamental human right'.

Among the first tasks embarked upon by members of the CDyPE was that of taking up contact again with affected rural groups who in the recent past had played a prominent role in the fight against contamination in the River Tunal. Oblivion, inertia, doubts and mistrust had first to be overcome, and this was done by going from one *ejido* to another, from the rural zone of the Durango munici-pality to the municipalities of Nombre de Dios and Mezquital. It was not easy to convince all rural producers of the need to support the research and the diagnostic study or of the necessity for 'intermediate solutions' to the contamination problem. For some, to speak of a need to investigate or evaluate levels of contamination and its effects was just a repetition of what had already – and unsuccessfully – been discussed, promised, and put into practice by

175

the relevant governmental organizations and business community. For others, the so-called 'intermediate solutions' were nothing more than a roughly disguised attempt to divert attention away from the main problem and to shield the principal culprit – the Celulósicos Centauro company – from the public eye. Yet others were of the opinion that the main beneficiary of any intermediate solutions would be precisely the business community, since 'the central objective of decontaminating the river would be pushed into the background in exchange for a series of insignificant works'. At times it was even necessary for members of the committee to overcome the suspicion that they themselves were working on behalf of the government or the business community, or that they were offering some newfangled product (in this case ecology) soon to be abandoned in favour of another.

Little by little, however, the issue took hold and convictions developed. The logic behind the mobilization and the proposed organizational scheme became obvious and convincing in the face of the magnitude of the problem and the professional and political soundness of the scheme's proponents. By the end of March 1989, the CDyPE had already shown itself capable of bringing together important nucleus groups of Registrillo, Arenal, 10 de Mayo, and Independencia y Libertad, all of which were situated less than 2 kilometres away from the cellulose plant and were therefore gravely affected by the contamination. Advances had also been made by the CDyPE in diagnosing and reporting on the contamination caused to the river by sewage from both the city and the cellulose plant,[12] and in work which later led to the formation of a Comité de Consejos Ecológicos (Ecological Advice Committee) in several *colonias*.

Parallel to these activities, the CDyPE opened up a dialogue with public institutions and the business community, and put forward a preliminary general proposal to find a way out of the impasse. Press conferences were initiated to explain objectives and proposals, and, with the support of the UPEZ, a direct channel of negotiation with the Subsecretaría de Política y Concertación Social de la Secretaría de Agricultura y Recursos Hidráulicos (SARH: Undersecretariat for Social Policy and Coordination of the Secretariat for Agricultural and Hydraulic Resources) was set up to begin discussions on 'intermediate solutions'.

Having arrived at the point where the campaign of action and the participation of the first popular groups was regarded as more or less consolidated, the time was ripe to begin public meetings with governmental and business institutions. It was proposed that the dialogue be established following a joint assessment tour of the affected zone. The visit took place on 26 May 1989 amidst extensive press coverage and was highly successful, resulting in the massive mobilization of the *ejidatarios* visited, some of whom aired their complaints in no uncertain terms. There were two further concrete triumphs resulting from this visit. One was the formation of the Comité de Vigilancia del CDyPE (CDyPE Vigilance Committee), formally created in the presence of officials and representatives of the business community to supervise the installation of the anti-contaminant system in the cellulose plant and the measures taken to cleanse urban sewage; and the second was the introduction of the Programa Ecológico Emergente (Emerging Ecological Programme), based directly on the demands and alternative solutions proposed by those affected.

As a result of these initiatives, the officials of SEDUE and the state government undertook to have the sewage treatment system in the city of Durango ready by 1989, and the business community for its part undertook to honour the agreement it had previously signed with SEDUE. For the CDyPE this meant a decisive step towards consolidation and the establishment of its standing among wide sectors of Durango society.

The Programa Ecológico Emergente focused on solving urgent problems caused by contamination, above all in relation to the productive activities and health of the affected inhabitants. In the case of agricultural production it proposed to find sources of clean water wherever they might be in order to avoid the continued watering of cattle in the contaminated river; advantage would be taken of such sources for developing other agricultural and fishing activities. As regards health, there was talk of building a specialized clinic in the affected zone. Further demands and proposals were added to the programme, such as the introduction of drinking water in deprived zones, the reparation and construction of sports grounds and schools, first-aid training courses, and the setting up of first-aid posts.

The reclamation of reservoirs for watering cattle in the zone was a matter which fell directly under the responsibility of the SARH, and the Programa Emergente thus specifically requested it to reclaim disused springs and construct small dams in the affected zones. The SARH reacted at first in an ambiguous and negative manner. Their initial proposal required that the applicants follow the due procedure and register themselves in the programme of public works through the normal channels, which necessarily implied that the works would be carried out at a tardy and indefinite date since the programme was already saturated and had a considerable backlog. Furthermore, the officials informed the applicants that the use of specialized machinery would cost around 25,000 pesos per hour, an extremely high sum considering that each community required more than one public work and that the wage of the machine operator (the lowest rate was 5,000 pesos per hour plus food and lodging) and the petrol and lubricants for the running of the machine also had to be paid. The estimated costs worked out extremely high (250,000 pesos a day) relative to the low income level of the affected rural workers.

The response of the CDyPE was simple and emphatic: its members rejected the idea of applying through the normal channels, which would make nonsense of the term 'emergency measures'; and instead of renting the required machinery they proposed having it on temporary loan in order to carry out the work themselves. The negative response of the SARH was not slow in coming, and the reaction was an immediate mobilization, culminating in a blockade of the streets giving access to the secretariat's central offices. Shortly afterwards negotiations were resumed with fully satisfactory results for the *ejidatarios*. An agreement was signed and endorsed by the Programa Emergente by which the bulldozers corresponding to the irrigation districts 01 and 03 were to be immediately transferred to the affected zones, where they would be run by the *ejidatarios* themselves. The Secretaría de Desarrollo Urbano y Obras Públicas del Estado (SDUOPE: Secretariat for Urban Development and Public Works of the State) undertook to assume all costs related to petrol and oil. The CDyPE then went on to form what was a key element in their strategy: the Unidades de Producción y Reserva Ecológica (UPyRE: Production and Ecological Reserve Units).

The UPyREs were conceived as an alternative intermediate solution to the core issue of providing pure water from reclaimed sources or dams in order to develop vegetable and animal farming on the basis of the logical exploitation, rationalization and harmonization of available natural resources. The process of creating UPyREs began as follows: producers would notify the relevant authorities about any source of running water they came across or knew to exist, and which was either blocked through disuse or simply not accumulated into a reservoir, or any dam capable of storing rain-water throughout the year. Technical studies for reclaiming the source or rebuilding the dam would then be carried out to provide a reservoir capable of holding the greatest possible volume of water. The machine operator would at once begin on the necessary excavation work to build the dam and reservoir. When the time came, fish were placed in some of the reservoirs in accordance with the relevant government authorities' specifications as regards species and number.

In the same way, fruit trees were planted wherever feasible with the aim of creating a protective belt to prevent water deposits and landslides and to improve the microclimate. The excess water from the sources and dams began to be used in a complementary form for the irrigation of garden produce, fruit farms, greenhouses, winter pastures, nursery gardens, and for improvement of natural pastures. The existence of clean water aided the setting up of an animal health programme to help cure livestock which had fallen ill as a result of drinking contaminated water. By September 1990 the CDyPE, together with the inhabitants of the zone, had succeeded in reclaiming 17 sources of water and had formed an equal number of UPyREs. Each unit had an average capacity enabling 1,000 cows to be watered, 3,000 fish to be kept, and 40 trees to be planted. A further 19 water sources are now being reclaimed to constitute a similar number of UPyREs, and there are ongoing experiments involving tree nurseries and greenhouses for the cultivation of flowers.

Once the functioning of the Programa Emergente was well consolidated, the CDyPE devoted itself to the task of promoting the Plan de Desarrollo Productivo y de Recuperación Ecológica (Plan for Productive Development and Ecological Recuperation), with the main objective of pinpointing specific causes of environmental depredation, finding both intermediate and final solutions for productive development, and restoring the ecosystem by means of the efforts and initiative of the inhabitants of the region. In launching the first phase of the plan the CDyPE consulted public opinion in all the *ejidos* and towns and villages involved in the struggle against contamination of the river so as to establish the demands and social and productive needs related to the phenomenon. This survey resulted in a huge variety of problems and schemes: the need for wells for drinking water and irrigation purposes, and various social and productive projects (from livestock, poultry and fruit farming sectors, recreational centres, public infrastructure works, etc.). Altogether, the plan initially embraced more than 60 activities geared to urban and rural needs. The CDyPE then went on to elaborate a medium-term financial plan envisaging the participation of the government and the business community.

It was calculated that roughly 6,000 million pesos would be needed to implement all these projects. The CDyPE proposed that the costs be apportioned in accordance with each sector's share of responsibility and ability to pay, as follows: 50 per cent by the business community, 35 per cent by official institutions, and 15 per cent (in money and in kind) by those directly involved. A preliminary testing of public opinion indicated support for going ahead with the plan; however, indecision soon arose as to who should administer the funds and how this should be done. The CDyPE was anxious to ensure that the funds be efficiently invested and that there be transparency and honesty in handling them. It was therefore proposed to form a trust, the evident advantage being that as a financial instrument it would be possible to build on the original resources through private tax-deductible contributions, as well as the fact that a legal entity of this nature best favoured the participation of the inhabitants of the zone in the management and control of its resources. In theory, the idea was to attract a significant initial amount of capital (2,000–3,000 million pesos) in order to be able to reap the benefits of sound investments on the financial market. At the same time this working capital would provide a permanent source of income to carry out the immediate priorities of the plan.

The formal structure of the trust envisaged the participation of federal institutions such as SEDUE and the SPP, state bodies such as SDUOPE and the Secretaría de Desarrollo Económico (Secretariat for Economic Development), and municipal bodies such as the municipal council of Nombre de Dios. The business community would be represented by the Grupo Industrial Durango (GID: Durango Industrial Group), and the social sector by the CDyPE and the UPEZ. Basic decisions would be arrived at by consensus, and a Consejo Técnico (Technical Council) was to be created to decide on priorities and types of productive and financial investment, on the basis of pre-established criteria.

The trust was formally set up on 11 July 1990, in the presence of the head of SEDUE, and was given the name proposed by the CDyPE: Fideicomiso Social de Desarrollo Productivo Regional

(Social Trust for Regional Productive Development). The first contributions were 100 million pesos from the business community, 30 million from the governor, and a symbolic million from the CDyPE in the name of the 'social sector'. At a later stage the Programa Nacional de Solidaridad (PRONASOL: National Solidarity Programme) contributed 300 million pesos, providing the base for the 50 per cent contribution from the government and the 15 per cent from the producers (accountable in terms of local labour and materials).

At the urban level, the CDyPE initiated a series of studies and activities at the beginning of 1990 aimed at providing more embracing and enduring solutions to the problem of contamination in the River Tunal, and more specifically, contamination of the urban environment through all causes. It was firmly believed that there could be no real and lasting solution to the contamination of the river if the measures already being put into practice were not backed up by others addressed to the indiscriminate felling of trees in the forests of the Sierras del Epazote and El Aguacate, where the hydraulic basin of the river is formed. In the city, efforts to deal with problems related to sewage disposal were broadened to cover the provision of water and drainage to the popular *colonias* and encouraging the exploitation of refuse by putting anything salvageable to practical use.

The approach adopted in the city has not differed greatly from that employed in rural areas; in fact, the standing and strong influence of the CDP in urban areas has made conditions even more favourable. Defence and Ecological Preservation Committees have already been formed in several CDP *colonias*, and their diagnoses and schemes are beginning to take form and direction. The process continues to combine popular mobilization with specific proposals brought to the attention of the public via conventional means of communication (press, television and radio), the use of wall slogans, and discussions and debates taking place in the sessions of the municipal council and the State Chamber of Deputies. The pattern of organization and participation which is gradually becoming consolidated entails the growing participation of the 'citizenship', thus both desectoralizing and desectarianizing the course of action and the orientation of the proposals.

CONCLUSIONS

In a genetic-structural sense it is possible to identify a series of clearly defined phases in the development of the CDP and the CDyPE, their social struggle, and their campaign to preserve the environment and its living and productive conditions.

1 The initial phase goes from the tenant struggle to reduce water rates to the forming of the División del Norte *colonia*, the first territorial base of the so-called Movimiento Urbano Popular (MUP: Popular Urban Movement) in Durango. At this point the organization was only at a very primitive stage, and depended largely on the presence of voluntary activists with a radical ideological outlook. These activists, whose maoism and radicalism dated from the time of the national student movement of 1968, were responsible for setting the basic course of the developments which followed.

2 The second phase came between 1973 and 1976, when a process of invasions took place culminating in the foundation of the Emiliano Zapata and Lucio Cabañas *colonias*. By this time the body which was later to become the CDP had already more or less consolidated its popular base, and the MUP had begun to show an unchallengeable social and political presence in the city. There thus emerged a new front with a foothold already established in the political life of the state and with a series of proposals – as yet incipient or still germinating – for introducing changes into urban life in Durango.

3 The third stage lasted from 1976 to 1979. During this period the MUP broke away from some of the 'organicist' ideas concerning its development in Durango City, as it became evident that there were alternative evolutionary paths to the growth of the initial activist popular nucleus (then made up basically of popular groups from within the División del Norte, Emiliano Zapata and Lucio Cabañas *colonias*). This was reflected in the fact that an initiative which was relatively external to the basic nucleus – i.e. the mounting discontent and agitation in other urban areas and sectors in the face of high electricity rates – led to the expansion of MUP's frontiers and precipitated the birth of a wider organization, namely the Comité de Defensa Popular General Francisco Villa de Durango (General Francisco Villa Popular Defence Committee of Durango).

4 The fourth phase developed from 1979–85. The initial conceptual

scheme and political plan of action were at their height: advances were being made in building up internal democracy at the rank and file level, in the *colonias* and the CDP itself, and in extending the bases of support and mobilization. This latter process acquired a new characteristic in passing from an exclusively territorial plane to the incorporation of organized sectors such as musicians and small business syndicates (called the *plataformeros* (platformists)) into the CDP. The cultural dimension now became an explicit ingredient in the process of formulating an 'alternative scheme'.

5 The fifth phase lasted from 1986–90, and constitutes the real, generalized move towards breaking new ground. A situation of relative isolation and a 'repressive siege' set up by the government sparked off the need to abandon the previous policy of abstention at the electoral level, and the decision was taken to participate in the electoral process. It soon became evident in the process of campaigning that the importance of participating in elections resided not so much in the tactical need to 'break the siege', but more in the fact that it provided invaluable opportunities for the work, expansion and development of the movement. It offered the possibility of opening channels of communication and links with other social sectors, it created a privileged forum for putting forward global political proposals and schemes for transforming society in Durango, it conceded varied coverage for political activities, and multiplied and extended dialogue at the local and national level with the government and other grassroots organizations and popular movements. This modified the pattern of social action and the conception of democracy, which now moved from discourse on democracy in urban quarters and *colonias* at the direct or assembly level to discourse on 'social democracy' on a wider social scale within Durango. The secret ballot and universal and direct suffrage were introduced for the internal election of party leaders, and the organizational structure of the CDP was improved. The feedback served to multiply efforts and open up new possibilities: the cultural room for action became enriched and more complex; new regional fronts were opened up and consolidated; organizational activities continued in the rural environment (with the formation of the Unión de Pueblos Emiliano Zapata, since the time of its foundation the sister organization of the CDP); and ecological activities and movements commenced.

In this period there was a repeat of the electoral experience, and in 1988–9 the party captured a federal deputy seat, two seats in municipal councils (Nombre de Dios and Súchil) and some ten *regidurías* in municipalities throughout the state. The CDP participated actively at the legislative and administrative level in the local chamber of deputies and the municipal councils, putting forward general political proposals and measures aimed at bringing about social, political and economic change within the state. The party also managed to reach an important reconciliatory agreement with the federal government whereby its past economic demands were conceded, thus enabling it to initiate a process of sustained development in productive activities.

It is within this context and this pioneering impulse – which led the CDP to tackle problems going beyond the scope of its grassroots organizations and *colonias* – that the environmental problems of Durango City and its rural surroundings became key motives for popular and civil concern and mobilization. The rhythm and timing of the 'ecological struggle' has been different in the countryside than in the city, but in both it has been a coordinated effort. From the outset, the approach chosen and then implemented reached out 'from society towards the defence of the environment and its resources', always returning to its social point of departure. In other words, there has never been an 'environmentalist' concept as such. This is evident on at least three levels: ecological protest has been a popular movement right from its origins; it has involved a search for intermediate solutions to concrete problems in the health and productive spheres; and it works towards reinforcing or 'complementing' social and popular organizational structures with diverse objectives.

The 'ecological struggle' is multisectoral and aims at impacting upon the collective awareness of the citizens of Durango, bringing government and business sectors together in the process; but neither in its conception nor its practical consequences has it lost its popular social base or fundamental axis. It is precisely for this reason that the move to internalize the ecological issue within the basic social sectors of the CDP, the UPEZ and various rural nuclei in the affected zone was so decisive, not only in the sense of developing an awareness, but more concretely in the form of providing cohesive organization. This led to the creation of councils in the *colonias* to take care of matters such as the disposal of sewage and other waste matter, the introduction of drainage, and refuse management. To this same end the underlying organizational structure in the countryside is based on Unidades de Producción y Defensa Ecológica (Production and Ecological Defence units) which play the dual role of technical units for planning and action, and social units for harmonizing and organizing efforts. This same idea is reflected in the approach to relating to and harmonizing with government and business sectors. The Fideicomiso Social de Desarrollo Productivo Regional was created on the firm condition that the popular elements and their technical and activist friends would not find themselves supplanted or subordinated by their 'associates' in the planning and decision-making process. Such has been the wager advanced against the yet greater and more resolute challenge of finding durable solutions to the problems generated by contamination in the Durango environment.

NOTES

1 Translated from the Spanish by Phyllis Johnson de Barrantes.
2 The Partido Nacional Revolucionario (PNR: National Revolutionary Party) came into being in 1929 on the initiative of the country's ruling sectors and was aimed at gathering the majority of existing political groupings together into one single organization under the direct control of the government. In 1938, during the six-year presidential term of Lázaro Cárdenas, it was renamed Partido de la Revolución Mexicana (PRM: Party of the Mexican Revolution) and acquired its final structure of 'organization of sectors' (workers, peasants and 'popular' elements). It was in 1946 that the government of Miguel Alemán decided to give it the name of Partido Revolucionario Institucional (PRI: Institutional Revolutionary Party).
3 The Partido Acción Nacional (PAN: National Action Party) came into being in 1939 as a reaction of some business sectors (particularly in the north of the country) against the popular and leftist politics of Lázaro Cárdenas. Since then it has been the second most important national party, though it was only since the 1970s that it began to have a significant electorate.
4 Particularly prominent were the struggles of the Liga Agraria Estatal de Mexicali (State Agrarian League of Mexicali) in Baja California, whose basic aims were the productive organization of the *ejidos* (smallholder collectives held under communal tenure in accordance with the agrarian laws; *ejidatarios* are rural workers with rights over these lands) and opposition to the North American cotton-spinning factories. They also struggled against land salinity caused by salt carried down the Colorado River from the United States, and for democracy by direct participation in the electoral process. Also important was the fight by fibre and candle producers in Coahuila, Nuevo León and Durango to raise the price of

185

their products and secure payment of the money owed them, and to protest against the monopoly and politics of the parastatal enterprise which bought and marketed their products. In the north east, peasants fought for the redistribution of land concentrated in a few hands, the influx of credit to their sector, the elimination of *coyotaje* and middlemen, and against corruption. The *ejidatarios* of La Laguna, whose organization had been built up during the years of the Cárdenas government, also began a struggle.

5 In this context, a *colonia* is a settlement originally established on invaded land. Although the legal standing of *colonias* is somewhat insecure, these settlements often develop stable and complex social and organizational governing structures.

6 In 1980 Durango was still considered a basically rural state, with 21.4 per cent of its GDP coming from the agricultural, forestry and fisheries sectors, as against 17 per cent for the industrial sector and 20.1 per cent for the manufacturing sector. Though not negligible, its industry is based fundamentally on small establishments and the processing of products directly related to the agricultural sector.

7 'One of the main currents of the radical marxist tendency of maoist leanings in Mexico responds to the anarchist, magonist and zapatist (supporters of Emiliano Zapata) tradition, to the influence of German Spartacist thought, and above all to the influence of the Chinese revolution in its maoist phase of popular war and cultural revolution. ... Maoism appeared in Mexico after the student revolution of 1968 in the form of groups such as Política Popular (Popular Politics) or the Frente Popular Independiente (Independent Popular Front), which, despite their different evolution, had in common the fact that they had been formed by university students and professors who decided to carry forward an ambitious project of 'going to the people', putting the priority on political work alongside the peasantry and *colonos* movements, with the idea that the revolution was a long and continuous stage-by-stage process which in the face of the inherent realities of the country (many of these being identified with the realities of pre-revolutionary China), was destined to progress from the countryside to the city, leaving in its wake 'liberated zones' under popular control which would in the future challenge the global power of the bourgeoisie' (Julio Moguel, *Los caminos de la izquierda*, Editorial Juan Pablos, Mexico, 1987, pp. 125, 126).

8 In May 1980 the Coordinadora Nacional Provisional del Movimiento Popular (CNMUP: Provisional National Coordinating Committee of the Popular Movement) was formed on the occasion of the Primer Encuentro Nacional de Colonias Populares (First National Congress of Popular Colonies), held in the town of Monterrey. In 1981 the organism changed its name to Coordinadora Nacional del Movimiento Urbano Popular (Conamup: National Coordinating Committee of the Popular Urban Movement), integrating over sixty popular organizations distributed throughout the main towns in the country. After the Third National Congress, which took place in May 1982, the Conamup spread to states such as Guerrero, Sinaloa, Mexico and Baja California, and

integrated more than a hundred regional social organizations. See Moguel 1987: 30, 31.

9 See Moguel 1990.

10 This comes from a popular Mexican saying: 'a río revuelto, ganancia de pescadores' (rough waters benefit the fishermen).

11 In Mexico, 'regidurías' are the offices corresponding to 'regidores' who are rather like municipal deputies. They form part of the 'cabildo', a meeting of municipal authorities and 'regidores', and play a management role.

12 In the face of the 'technical' opinion of SEDUE, vastly underestimating the level of contamination in the River Tunal, the CDyPE undertook a multidisciplinary investigation which came up with irrefutable results to the contrary. On the basis of fourteen parameters, the existence of faecal coliform organisms in quantities significantly higher than those permitted by the SEDUE was clearly demonstrated. The results were similar in the case of the heavy metals lead and iron. As regards the colour of the water, all the samples with the exception of one were darker than the permitted range, the situation being critical in the actual discharge zones. The level of sedimented solids was excessive in the *ejidos* 20 de Noviembre and El Mezquital and the level of suspended solids totally exceeded the permitted values in all the samples. The chemical oxygen requirement, a vital parameter, showed irregularities in more than half the samples. The case was similar, though to a lesser extent, with the biochemical oxygen requirement. A comparison of these two latter parameters showed greater values for the chemical oxygen requirement, indicating the preponderance of inorganic contaminants, i.e. those of industrial origin, which require a larger quantity of oxygen to render them less harmful. All samples surpassed the permitted levels of greases and oils.

REFERENCES

Moguel, Julio (1987) *Los caminos de la izquierda*, Mexico City: Editorial Juan Pablos.

Moguel, Julio (1990) 'Las coordinadoras de masas: una aproximación a la crítica en sus diez años de vida', *Hojas* No. 1.

8

STRATEGIES FOR AUTOCHTHONOUS DEVELOPMENT: TWO INITIATIVES IN RURAL OAXACA, MEXICO

Jutta Blauert and Marta Guidi[1]

INTRODUCTION

This chapter presents two initiatives taken within indigenous communities in southern Mexico; initiatives designed to ensure the rural poor's control over their physical and social resources. The different socio-political dynamics within and surrounding these communities, the environmental context, and the historical experiences with state-led development programmes are considered as background to the different forms of activism adopted. The studies show how a threat to the natural resource base and local livelihoods, either by direct outside intervention or by environmental deterioration, can trigger concerted community action. In addition, the immediate need to ensure daily income may move the rural poor to initiate their own production projects, which subsequently extend to encompass environmental issues and political action concerning control over natural resources.

The contrasting cases are described to depict some aspects of the diversity which characterizes local-level activism and local knowledge systems, for it is this diversity upon which sustainable policies for resource management must ultimately be based. Both initiatives in Oaxaca demonstrate the centrality of indigenous peoples' grievances over economic development and political exclusion and repression, and their demands for power to define the terms of development projects. Organization, therefore, is frequently centred on ethnic identity and demands for cultural re-evaluation.

This chapter will argue that:

188

1 In Oaxaca, locally based initiatives are, by force of diversity in historical developments and contemporary human and physical environments, varied and highly locality-specific. In addition, each has developed out of an idiosyncratic social dynamic that has led to different political strategies and social networks being employed. Yet they share certain similarities in terms of wider objectives, including increasing emphasis on the cultural sphere and the now more frequent use of environmentalist discourse.

2 These local initiatives have most commonly sprung up in response to one or a series of unsuccessful or even threatening state-led rural development programmes. The specific actions undertaken are frequently determined by the local and regional political contexts, as well as the existing degree of ethnic assertiveness.

3 The two cases will also demonstrate that just as 'sustainable development' is not sufficient where it ignores the local socio-cultural and physical environments (we shall, therefore, employ the term 'autochthonous' development[2] instead), so the provision of means to encourage local 'participation' in resource management is insufficient where local-level *control* over natural resources does not exist. For these rural people, mostly small-scale indigenous farmers, participation in the management of their resources is insufficient. What is essential is control over their resource base in the first instance, as well as local empowerment and the assurance that they can play a role in the broader decision-making process regarding how local and regional resources and production are to be used and allocated.

Furthermore, if the lessons from past failed rural development and environmental policies are to be learned – and applied – then 'participatory action research', 'participatory technology development' and 'participatory projects' need to incorporate an understanding of the value of local knowledge. Much activism relies on just that knowledge, and the rural poor have based their livelihood strategies on it. Local knowledge, and particularly indigenous knowledge, in both the agroecological and the sociopolitical spheres, needs to be re-valued at all levels of policy making and implementation/extension. It is this knowledge which can enable the rural poor to be *actors* in sustainable development.[3] Activists, project members or NGO members, on the other hand, must also acknowledge that, in certain deteriorated physical and economic environments, some traditional knowledge systems on their own can no longer provide all the answers.

CHALLENGES TO SUSTAINABLE RESOURCE MANAGEMENT IN INDIGENOUS COMMUNITIES IN OAXACA

The southern Mexican state of Oaxaca is commonly depicted as a region of largely indigenous peoples, relying for their livelihood primarily on subsistence rainfed agriculture. The contrasting agro-ecological regions, which include humid, sub-humid, temperate and cold mountainous climates, from Pacific coastal areas to the central highland plateaux and a number of very rugged and eroded mountain ranges, provide a picture of varied but generally poor natural and socio-economic environments. Oaxaca is, in effect, one of the poorest and most eroded states in Mexico, with one of the highest rates of seasonal and permanent outmigration of agricultural and domestic labourers – mostly small-scale peasant farmers. The state of Oaxaca is divided into 570 municipal authorities, and is thus characterized by fragmented administration and poor local finance while regional and federal governments control the highly centralized decision-making processes.

Only 14 per cent of the area of Oaxaca is arable land, 42 per cent being under tree cover, and 23 per cent of the land being totally eroded. There is some coffee, sesame, fruit and livestock production, but of the arable land 78 per cent is rainfed agriculture, and 80 per cent is under the basic crops of maize, beans and wheat. Yet production does not cover demand: 72 per cent of maize consumed is supplied by production in the region, but only 30 per cent of wheat can be supplied internally. Farm size averages officially 4–5 hectares, but minifundios of 1–2 hectares predominate in the sierras. Three-quarters of the land is held by indigenous communities or as *ejido* land, but these lands are mostly rainfed lands or unimproved pasture and woodlands; only 0.5 per cent of the lands affected by agrarian reform are irrigated or humid lands. Over 80 per cent of the rural sector have an annual income of under £120, and Oaxaca's officially set minimum wage is one of the lowest in Mexico. Forty-four per cent of its people are indigenous people, from 16 different ethnic groups (Beaucage 1988: 197; SPP/INEGI 1985; SEDUE 1986: 53; Reina and Sánchez-Cortes 1988; Mexico/Oaxaca 1982).

In the regions of our case studies, the Mixteca Alta and the Chimalapas/Lowland Mixe (see Figure 8.1), as in Mexico in general, rural development planning – governed by agricultural and economic policy as well as by the 'integrationist' ideology of indigenist policy – so far has provided no effective, equitable or sustainable alternative to environmental deterioration and falling living standards.[4] Instead, 'development' plans are frequently the very cause of environmental degradation and of socio-political conflicts. Outsiders, central planning bureaucracies, political parties and local institutions – themselves governed by international influences – are only too often inflexible, and rooted in different economic rationalities, political and social interests to those of local ethnic communities. They also tend to ignore indigenous technical knowledge and the socio-cultural environment.

Analysis of agricultural and regional development policies, such as the Mexican Food System (SAM) and the Integrated Rural Development Programme (PIDER), has indicated that, despite changing rhetoric, little progress has been made in the last two decades. There still predominates a vertical concept of production and planning (Wall 1982; Mexico/Oaxaca 1984). Changes in planning and development terminology have not altered underlying concepts and objectives, nor have they truly changed extension attitudes and practices. Peasant farmers still rarely have an understanding of any given policy or programme and are therefore unable, even if they were to be invited, to evaluate its contents and objectives before implementation.

The policies that have had the most visible long-term effect on the Oaxacan environment and, above all, on its indigenous people and their land have been the irrigation projects, logging concessions and colonization programmes of the state. Over the last three decades state intervention to create irrigation infrastructure for the areas in the state with the best agricultural lands has been drastic and plagued with conflicts.[5] The subsequent emergence of large sugar plantations and new irrigated lands stimulated the immigration of many settlers from lowland mestizo areas, who were encouraged by the state to settle on lands formerly part of indigenous communal territory. The frontier of the humid tropics in southern Mexico was thus extended, leading to further environmental degradation and social disintegration.

In the 1950s the government also gave logging licences for the Tuxtepec/Sierra Juárez area, on the border with Veracruz state, to

191

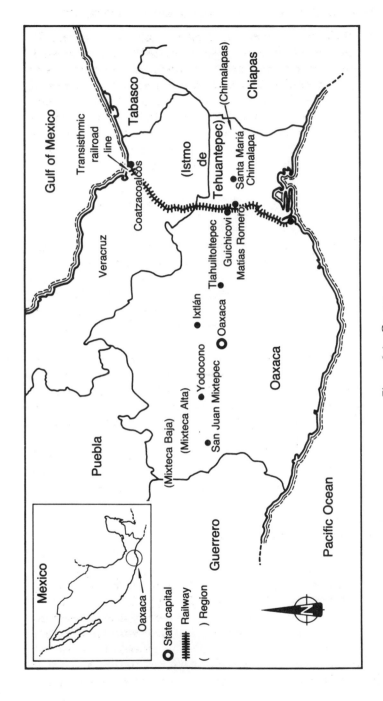

Figure 8.1 Oaxaca

Source: *The International Atlas* (1991) Chicago: Rand McNally & Company

a number of private and parastatal lumber companies to ensure the national supply of cellulose and paper. The resulting environmental, economic and political problems, the deterioration of livelihoods of indigenous communities still cause extensive conflict in the region, particularly in the Mixe areas. A range of state-led development programmes followed in the 1970s and 1980s (decades of the 'integrated' rural development planning) without improving the situation of Oaxaca's poor, nor addressing the issues of people's participation in decision-making or resource management.

THE SOCIO-POLITICAL CONTEXT

The agricultural crisis in the 1970s, the economic crisis since 1982 and the new modernization paradigm that governed regional development planning have thus led to stresses and dysfunctions at different levels. As a result of these processes, and of the political crises in Oaxaca since the early 1970s, the dynamics of a variety of social movements have influenced the emergence of indigenous peasants' actions and organizations such as those described below.[6]

Of particular influence in the region was the students' movement after 1968, which, together with transport workers and traders, brought down the Oaxaca government in 1970. Under an increasing political polarization of private and public sectors in the state, and land conflict and violent repression in the mid 1970s, dissident groups in 1972 formed the 'Worker-peasant-student Coalition of Oaxaca' (COCEO), and the 'Worker-peasant-student Coalition of the Istmo' (COCEI) in 1973. Together with the 'Independent Peasant Front' (FCI) (formed in 1976) and regional trade unions and ejido organizations they succeeded again in deposing the governor in 1977.[7]

However, the conflicting ideological and strategic positions taken by political parties in these movements, and some errors committed by teachers and students in sierra indigenous communities, frequently led to increasing violence and divisions in the villages and movements rather than the solution of the original problems. From 1980 onwards, therefore, specifically indigenous groups began to assert themselves at local and regional level according to their own organizational structures and with their own leaders. In parallel to national developments, indigenous organizations established closer links with one another, and many joined the independent national peasant confederation 'Coordinadora Nacional Plan de Ayala' (CNPA), founded in 1979.

The links established and joint actions undertaken by national federations of other popular movements were reflected by similar expanding links between movements from different social sectors in Oaxaca. At the regional level, the experience of the COCEI and the independent teachers' movement, which includes many indigenous teachers, in the early 1980s provided a new focus and support for indigenous movements. Throughout Oaxaca participation by dissident organizations in local elections – usually only possible in alliance with left-wing parties – became widespread.[8]

But at the village and municipal levels, few indigenous communities and organizations considered it appropriate to enter into a close, formal alliance with a political party for a joint electoral campaign, such as that undertaken by the COCEI. Ideologies differed, or it was feared that the electoral strategy would attract further repression. Nonetheless, the support of sympathetic opposition parties and popular organizations was constantly sought by the indigenous movements.

Given the history of the tightly organized corporate state in Mexico, the role of publicly tolerated but politically independent institutions such as NGOs has been crucial in the new focus given to rural development strategies by indigenous movements. There are a number of non-governmental organizations (NGOs) and 'promotion organizations' (OPs or grassroots support organizations) based in Oaxaca City, most of them working with indigenous peasant communities and their organizations in production, health, education and organization. Their support to communities has been important in many ways, but they and individual researcher-activists also have been criticized on several occasions by indigenous people for taking on too much of a leading role and manipulating funding sources.

The emerging organizations of indigenous peasants struggle, then, for a series of objectives, each relating to environmental concerns:[9] for respect of human rights; for control over natural resources and land; for democratic local elections, for state support in the provision of public services, production and marketing; and frequently for defence against ranchers and logging companies. 'Environmental' issues, therefore, are rarely a primary or sole concern of indigenous peasants. But such issues are increasingly found in the public discourse, as if in response to government

194

rhetoric on sustainable development. On the other hand, the defence of natural resources, which are economically essential and heavily imbued with cultural values, and of the right to determine their use, has always been intricately linked to struggle against corrupt authorities.

Contemporary indigenous peasants and their communities are affected by a far wider range of experience, tools and external influences than were previous generations (mass media and communication networks being one example). Ideological and practical expressions of resistance and adaptation thus show a variety of strategies and tactics, according to the degree to which economic systems and cultural influences have or have not destroyed communal and ethnic identity and self-provisioning. The initiatives presented below reflect only some of the indigenous movements occurring within the state, but they allow us to consider the similarities and differences between at least two autochthonous approaches and their dynamic developments.

THE MIXTEPEC PROJECT: PARTICIPATORY COMMUNITY AND AGROECOLOGICAL DEVELOPMENT

The regional background: the Mixteca

The Mixteca is located in the north-western part of Oaxaca. The region spans a variety of agroecological areas and is characterized by a highly abrupt topography, rapidly advancing erosion, and infra-subsistence production of the traditional food staples of maize, beans and squash. The Mixteca highlands cover altitudes between 1,300 and 2,600 metres; climatic conditions of temperate highland areas and semi-arid upland chaparral predominate. Rainfall averages 500–700 millimetres, and generally poor soils and steep slopes further worsen agroecological conditions.

Today only some 6–10 per cent of the region is cultivated; 93 per cent of it is under rainfed cultivation with rainfed smallholdings of 1.5 hectares prevailing. A quarter of the area is totally eroded, and another 20 per cent in advanced stages of erosion. Farming systems in this region are those of predominantly small-scale subsistence farmers growing maize, bean and squash. Farmers encounter problems of insufficient (or untimely) capital/credit, coupled with high input costs; insufficient access to land, insecurity of land tenure, conflict over community borders and land invasions by ranchers; climatic risks (frosts, drought, untimely rains, falling water-tables); seasonality of labour (migration); soil erosion/low yields (maize yields are about one quarter to one half of the national averages); and lack of any control over the agricultural market.

The roots of these and other contemporary problems in the region are to be found in historical processes, but the pace of change and the range of external influences and internal transformations have increased drastically over the last decades. An analysis of the history of the natural and socio-economic environment of indigenous communities in the Mixteca has shown that the origins of environmental degradation go back decades, but that more recent changing production methods and economic relations have led to cyclical patterns of a more intensive, and spatially extensive, degradation across the region (Blauert 1990). By the time that new crops, agricultural technology and economic forces of the twentieth century began to cause a dramatic escalation in the impoverishment of the resource base of indigenous communities, the heritage of historical conflicts had lain the foundations of many contemporary political and economic conflicts.

Relations between villages and their external world are in most cases characterized by vertical dependence and horizontal conflict. Today the Mixteca is internationalized less by the extraction of its physical and human resources within the region than by the extraction of its labour outside the region, and by production and consumption patterns introduced from outside (Kearney 1986). Remittances from migrants exceed the value of agricultural production in the region.

Negative experiences with agricultural and rural development policies, and deepened tension over agrarian issues since the middle of this century, have sharpened the conflict between indigenous communities, private enterprise and the state. From the early twentieth century (and particularly since the 1950s and 1960s) some of the actors in this 'game' have changed. The state has taken on a more directly mediating role, and teachers and returned migrants have established themselves as highly influential groups. Inter-community land conflicts continue to be widespread due to inappropriate, non-existing or uncompleted titling by state agencies.

This situation is worsened where mestizo entrepreneurs, backed by gunmen and often with the help of municipal/district mestizo authorities, invade land for farming or ranching. Conflicts between such invaders and villagers have led to increasing physical violence over the last two decades. Frequently, mestizo *caciques* (local political strongmen) arrange to 'sell' communal land to other *caciques* through their access to municipal authorities and sometimes with the aid of compliant *bienes comunales*[10] officers. Where peasants organize to demand participation in local politics and development, authorities commonly support capitalist interests through repression and instigation of internal conflicts, and fraudulent management of legal documents; fraudulent elections and inaccurate electoral lists serve as a backdrop to the repressive use of army and secret police forces by the authorities.

PLANNING IN THE MIXTECA

State development plans date back as far as 1935 with programmes meant to 'alleviate poverty' by promoting forest exploitation, fish and silk production and the development of indigenous crops, such as agave, into cash crops (Méndez Aquino 1985). During the 1930s and 1940s, the national colonization programme for the humid tropics also affected the Mixteca. These mostly unsuccessful programmes promoted expansion in road construction, private logging and mining activities, and thus increased deforestation and migration.

In subsequent years, the National Indigenist Institute (INI) gained a foothold in the region, but its programmes only took on some substance after the creation in 1960 of the Riverbasin Commission, Comisión del Río Balsas (CRB). The CRB attempted to coordinate agencies' programmes in the region to promote an integrated economic development, centring its actions, however, on infrastructure development such as irrigation, electricity supply and road construction. The arrival of the PIDER programme in the Mixteca in 1974 offered additional funding for 'integrated' development projects (predominantly production and social services infrastructure). The commitment of some officials to such aims, however, did not suffice to substantially change the prevailing practice of anti-participationist intervention in the region.

THE SAN JUAN MIXTEPEC PROJECT: 'RECOVERING WHAT'S OURS'

The initial focus of this independent project was one of soil conservation, reforestation, and sustainable agricultural production in communities of small-scale farmers working within a temperate and semi-arid highland environment. The longer term objective is to re-strengthen and re-claim local cultural values, ethnicity and the organizational capacity of the community so as to facilitate self-determined and community-led development.

The project emerged from the intention of placing emphasis on the concerns and experiences of the Mixtepecos regarding their environmental and production problems, and of combining traditional agricultural knowledge systems with alternative technologies that are simple, cheap and easily adapted to the local context. Initiated as a small-scale project for environmental improvement, increasing numbers of peasant farmers participate in the work, and the working spheres covered by the project have diversified.

The context

The municipality of San Juan Mixtepec, which covers 35 villages, lies in the Mixteca Alta, with a great part of its lands lying on steep and eroded slopes. Water in general is scarce. The political borders of the municipality coincide with the historical territory of a Mixtec community which today has some 14,000 inhabitants who are linked through kinship, linguistic, religious and political contacts and through the exchange of goods and services.

For the indigenous peasants of Mixtepec, as for almost all of those in Latin America, the 'quality of life' is determined and constrained by a disadvantageous insertion in the market economy and the socio-political structures of the Mexican nation-state. As a result, its capacity to satisfy its basic needs depends to a large extent on the quality of its environment, on its production strategies and the multiple activities taking place within the family group.

To talk of environment in Mixtepec today is to talk of deterioration: of mountains eroded by deforestation, burning of vegetation and overgrazing; eroded arable lands, with a tillable surface that rarely surpasses 10 centimetres; lands exhausted by the intensive cultivation and excessive use of chemical fertilizers; rivers whose current decreases each year, and small lakes, bogs and springs which are slowly disappearing in the long droughts.

The peasants of Mixtepec, conscious of the impoverishment of their living standards, have tried for years to find ways of countering falling production and lack of water, especially by participating in governmental 'development' projects. However, given the political

198

conflicts within the community and the operational characteristics of governmental agencies, the planned projects have never been fully implemented.

Their negative experiences have caused the Mixtepecos to identify their situation as a problem whose solution appears to depend on power spheres outside the control of the community. In general, contact with national society has implied for the Mixtepecos the strengthening of a structural inequality that keeps Indian groups removed from the centres of decision-making (see Guidi 1988a). The history of their relation with official institutions confirms this assessment: not only have they been marginalized from all participation in relation to the basic service infrastructure of their municipality, but also from any decision-making processes concerning the development of their community. The Mixtepecos, therefore, were more or less forced to accept the technical solutions offered – which did not consider community perceptions of problems, nor the internal conflicts of the group or its own conception of agricultural work. In the end, such technical solutions, themselves excessively subject to budgetary problems and immediate political demands, always ended in failure or remained incomplete. The response of the communities was to develop an attitude of apathy and scepticism.

The history of failed 'magical' solutions led to a sharp increase in illegal migration to the United States. Instead of improving the situation, however, migration has increased the communities' dependence on consumption of imported goods, particularly of status symbol goods; it has implied an additional workload for women in the community; and has contributed to worsening environmental deterioration because it led, for example, to a more intensive land use and a higher application of chemical fertilizers.[11]

In this context, and as a result of the worries shared by some peasant farmers, the new local authorities, some politically active groups of the municipality, and two 'outsider' researchers working in the villages at the time, there emerged in late 1986 the idea of elaborating jointly a more structured autochthonous response to the municipality's problems. This 'project' was to combine attention to environmental and production problems with the rehabilitation of those cultural aspects most directly linked to communal identity.

Initiating the project

However, given the history of local experience with 'development projects', the transition from a 'joint perception of a problem' to the 'joint implementation of alternatives' was a long process. It implied at first many months of long informal discussions between those interested, the collection of different proposals from within the villages and contact with other, similar projects to evaluate experiences and the viability of the suggested 'new-old' technologies. Although most people coincided in their assessment of existing problems, the attitude to the possibility of initiating joint work varied in different villages. Previous attempts to set up a cooperative for basic food supply and harvest sales had failed, for instance, due to internal factionalism and inexperience in organization; and the communal, but later largely individual, work with fish tanks, vegetable production and goat husbandry had, because of the projects' failure, left people with a bitter attitude towards any future joint efforts. The scepticism of the upland communities also reflected their conflictive relation with the central village and the sharp confrontations caused by the powerful position of the *caciques*,[12] who in the summer of 1986 still controlled the municipal authorities and the *comisariado de bienes comunales*.

A decisive change occurred with the election of the new authorities in September 1986. From the beginning, the joint proposal had caught the interest of two local groups, interested in recovering control of the municipal authorities: the Comité de Maestros (an organization of local teachers with a background of training in the city of Oaxaca, concerned about the defence of communal rights, land, natural resources, health, education, etc.) and the Comité Voluntario (a group of young migrants, all of them having been undocumented agricultural workers in the south-west United States, having links to migrants' organizations in California and in northern Mexico, and particularly conscious, through their own migratory experience, of the importance of re-strengthening communal self-identity and to maintain group cohesion). The activism of the Comité Voluntario in the upland communities had always been intense and the group rapidly obtained support from the majority of inhabitants. As a result of their activities, in July 1986 a large communal assembly was held which elected a new municipal presidency,[13] which was in charge of a team made up of teachers and migrants.

This new local authority was especially concerned about the environmental degradation and its effect on the socio-economic situation of the municipality. The team's objective was to protect natural resources from outside exploitation[14] and from inappropriate local use, while at the same time looking for alternatives which would benefit the poor families of the highland villages. The change in local authorities was, thus, decisive in accelerating the implementation of the project and its dissemination amongst the communities.

In this process, the visit in November 1986 of the new municipal authorities and some other members of the community to the project of conservation and low external input agriculture coordinated by CETAMEX in Yodocono, a Mixtec village with similar agroecological problems to Mixtepec, was very important.[15] There the Mixtepecos could observe, in practice, and discuss with other Mixtec peasant farmers the possibility of finding useful measures to contain erosion and to improve the productivity of their lands. After the possibility of implementing new proposals had been demonstrated by the visit to the CETAMEX project, the showing of films, and visits to different local agroecological zones and plots, at the beginning of 1988 some peasants and the new authorities decided to constitute the Comité Pro Proyecto Mixtepec, which assisted the small team in charge of coordinating the implementation of the proposals.

In this first stage the project concentrated primarily on trials of simple alternative technologies designed to solve environmental problems and improve the productivity of arable lands. The concept was one of a project made up of a number of related smaller projects, concentrating on soil conservation, water harvesting, agriculture and reforestation. At the same time, work was slowly begun on aspects of health, on workshops and discussion groups about forms of community organization, and on assisting with problems concerning land tenure issues.

Given the characteristics of the community, and in agreement with the Comité, it was decided to initiate work only with a few farmers and in few villages, but to attend to each case in an intensive way in order to guarantee a deeper understanding of the issues and a better knowledge of the techniques involved. In this way, the dissemination of proposals coming from the grassroots, the farmers themselves, was also guaranteed.[16] The project was initiated formally in March 1988, operating only in two communities of the

municipality; by July 1990 different activities were being undertaken in 20 localities, involving an average of 250 families across the
whole area. In most cases it has been villages informed about the
project work by their neighbours which have approached the
project members to arrange for joint activities.

Methodologically, the project insists on the direct participation
of those interested, as much in the diagnosis of situations as in the
selection of alternative appropriate techniques and their use, and in
the periodic evaluation of work carried out. A fundamental principle of the project is respect for the capacity by farmers to make
decisions, for their organizational forms and their own conception
of time. For the same reason, there has always been an insistence
on maintaining only a very small team of 'outsiders'[17] so as to
change as little as possible the communal arrangements.

At the same time, in all of the activities the informal emphasis is
placed on re-strengthening the self-perception of the Mixtepecos as
a group. The aim is to ensure that when selecting and applying new-
old technologies, an incentive is created for the recuperation of a
socially affirmative self-understanding and the development of an
organizational capacity which facilitates the step towards communal
self-management.

The activities of the project

Originally conceived as the sum of various small projects primarily
focusing on environmental deterioration, today the project works
in three basically interrelated areas: conservation and improvement
of production and the environmental situation; training of health
promoters; and the setting up of a producers' cooperative.

The implementation of alternatives to environmental deterioration continues to be a priority action, given its direct effect on the
quality of life of the Mixtepecos. In this respect reforestation has
been carried out in 8 communities already. Native species, especially
pine and leucaena, are preferred, although in some cases the
communities have combined these with fruit trees. The project
provides technical assistance and economic support for the purchase
and transport of the trees; the work of selecting and preparing the
plots, the sowing, fencing and general care is carried out in the form
of communal collective labour schemes (*tequio*). There have also
been some 500 fruit trees planted by families from 8 communities.
Nurseries have been established in 4 villages for horticultural

produce, fruit trees and seedlings. The nurseries also fulfil an important educational purpose, and serve to prove the viability of introducing new (but traditional, indigenous) crops, such as amaranth, and contribute to the avoidance of a dependence on governmental institutions. Thirty-two families from 6 different communities have also begun to establish small family nurseries on their land. In all cases, the nurseries are designed to grow horticultural crops and fruit trees or trees for reforestation; they also serve to encourage the use of organic fertilizers and biological pest control.

With fruit and vegetables from the nurseries, communal gardens have been established in 5 communities. In all cases, the tasks are carried out by *tequio* work, and each community decides independently on the distribution and use of the produce. However, the most intensive work is at family level: almost 80 families of 12 communities already have fruit and vegetable gardens in which organic fertilizer use and biological pest control are being encouraged. As spaces used primarily for individual and group experimenting, the work in gardens stimulates confidence in local agricultural knowledge. Tended mostly by women, the gardens and nurseries have served also to increase female presence in discussion meetings and in decision-making.[18]

Much more complicated, however, is the labour-intensive building of terraces (contour bench terraces with ditches and bund strengthening hedges), given the lack of male labour in the dry season due to migration.[19] In the first two years of the project bund-ditches have only been built in nine plots. However, the difference in the productivity of the terraced plots is so well known (especially in the more humid lower parts) that many neighbours of the 'beneficiaries' are planning the construction of bund-ditches for the coming winter.

After a decision by the assembly, primarily due to insistence by the women, since late 1989 work has begun in assisting the development of beekeeping in two communities. Beekeeping is becoming primarily a 'female' activity, not only because of women's ability for this work but especially because of the space which the women have won in the decision-making process.

Women have been continuously gaining a political space also within the cooperative Yosonovico. The formation of a cooperative society to serve the farmers linked to the different activities of the project was a fundamental concern since the beginning. Given the characteristics of the social relations in Mixtepec, and considering the many previous failures of cooperatives, it was decided that encouragement of and assistance to the formal setting up of a cooperative society was to be handled in parallel to the implementation of a programme of workshop courses in training and discussion. With this vision, since the middle of 1988 informal discussion meetings have been combined with intermittent courses.[20] In June 1989 the first formal constituting meeting was held of the Yosonovico Cooperative of food producers, made up of 11 local farmers (8 men and 3 women) linked to the project. The cooperative tries to produce, on a small scale, agricultural produce for local consumption and for commercialization in regional markets. Members contribute minimal quotas, a plot (as a loan) and their work. Today, the plots have been sown with vegetables, fruits, alfalfa and beans. The cooperative also produces honey for local sale.

Together with operation, production and marketing, the project assists with training the members of the cooperative through weekly meetings of discussion and planning, cycles of courses and specialized assistance. Since April 1989 two members have been participating in a number of regional meetings of cooperatives from south-eastern Mexico, so as to deepen their knowledge and to exchange experiences with other farmers' groups in the country. The meeting with other peasant farmers with different levels of organization has been an important element of reaffirmation.

In the area of health, the project is preparing the conditions which are to allow the implementation of an integrated health project. As a first step, the project has concentrated on the training of local health promoters through courses, workshops and specific assistance,[21] and in the joint recovery of communal knowledge of medicinal plants. Slowly, this knowledge, like that of traditional curative practices, is being reconstructed. The promoters are trained to diagnose and provide basic treatment for the most common illnesses in the municipality; they are beginning to prepare their own applications and to develop work on four gardens for medicinal plants. Their periodic participation in regional meetings of indigenous doctors and alternative health practitioners serves not only to enrich their training and to foment the exchange of knowledge, but also to strengthen their self-confidence.

Related to all these activities, and especially to the wider issue of community organization, are the problems of land conflicts, dealings with bureaucracy and general legal issues. Courses are being given to explain different legal procedures and rights to community members and especially to members of the *comisariado de bienes comunales*.

Some initial ideas have had to be adapted as circumstances changed: migration is getting more intensive each day in the villages and affects the development of all work areas; armed confrontations over land conflicts have occurred, occupying the whole attention of the community; there has been interference from a governmental project more concerned with co-opting the farmers politically. Existing as an independent project which maintains more or less close relations with some NGOs and with governmental agencies, but which does not subordinate itself to the interests of any group or party, has implied a major battle – for instance in obtaining outside financial support. However, in spite of these problems – and as a result of having remained a small project, independent and concentrated in the internal relations with the community – an increasing degree of participation has been achieved, not only in terms of numbers of participating communities (20 in 2 years) but also in terms of the participation of farmers and authorities in the different activities.

At the same time, the organizational function of the project staff is slowly being reduced, and farmers are taking over more responsibilities. Through collaborative work and discussion at assemblies and workshops, participants ascertain that it is they who can and do take the decisions. The work involved gives little incentive to richer farmers to participate (no machinery like tractors is offered, nor are traditional cash crops used). The aim of the project – to work primarily with resource-poor farmers and especially with women – is thus largely put into practice.

Already the initial stage of promotion and organization has been left behind, having required a major input in activism on behalf of outside actors. On the other hand, the project is still in its early stages. Its underlying principle – that it should evolve in accordance with the initiative of the farmers – can make it appear to the eyes of a conventional observer as a 'small' project, difficult to evaluate in terms of conventional cost–benefit indicators. But it was, and is, exactly a central objective of the scheme to remain a 'small' project for a long time, in order to demonstrate the viability of small development projects which do not imply high costs or sophisticated machinery, which avoid as far as possible the invasion of 'specialists' in the communities, which involve farmers without pressure, and which are feasible for future expansion – albeit slowly – from within the community. The important point in this case is that the increase in participation, discussion and assertiveness, the success in the practical work, and the form in which the self-dissemination of the proposals has been achieved, all have shown that this type of initiative can help indigenous farmers to re-strengthen their social identity, and to gain self-confidence in the re-evaluation of their role in the socio-economic and political systems.

A state-independent, small-farmer-oriented project, with few active participants to start with, but with many cautiously supportive onlookers, is slowly developing into a self-managed project in which more and more farmers participate. For the time being, small and slowly emerging, but solidly founded, successes in different areas can provide more tangible results, as well as foundations for future constructions of alternatives.

THE UCIZONI: WORKING FOR THE FOREST AND ITS PEOPLE

The case of the 'Union of Indigenous Communities of the Northern Isthmus' (UCIZONI) presents us with a different form of activism around environmental issues, with spheres of action much broader than those of the Mixtepec project. The explanation for this in part is to be found in the different environmental situation of the lowland Mixe, but is also largely rooted in differing socio-political developments, especially changes in resource use as a consequence of state intervention and the economic interests of the private sector. The geographic situation of the villages where the UCIZONI is active also plays a role, being located along the transisthmic railway line and roads.

The Mixe reflects many of the developments threatening livelihoods and ecosystems in the Mixteca: environmental degradation with parallel changes in farming and knowledge systems and deterioration of livelihoods, threats to cultural identity and natural resources as well as increasing internal class differentiation, coupled with a rising dependence on external inputs have led to political and socio-economic crises in (and between) the villages.

The regional background

The Mixe region, opening onto the Istmo, is as diverse as the Mixteca, being commonly divided into the highland Mixe Alta and the lowland Mixe Baja reaching eastward towards the border with Chiapas and the Zoque area in the Chimalapas. To the south it borders on the coastal plain of Zapotec villages, and to the north on the states of Veracruz and Tabasco, having as neighbours the Chinantec and Mazatec peoples. The temperate and cold highlands are characterized by maize subsistence production and *minifundismo* and high outmigration, while sub-humid and tropical lowlands with areas of tropical rainforest in the most eastern areas of the Isthmus allow coffee and fruit production. The lowlands have produced coffee for some time now, but logging for tropical hardwood and cattle pasture is rapidly proving a greater disturbance to ecosystems than subsistence agriculture and small-scale coffee production by the indigenous and settler population had been.

In the lowlands, land is held mostly by *ejidos* and irrigation facilities are frequent; sale of plots and land speculation is therefore common. The promotion of cattle ranching by agricultural banks in the region has led to problems of indebtedness among the population. Some villages, in the attempt to obtain some income have either leased some of their forested lands to private lumber companies – receiving little of the value of timber extracted and the value added through processing – or have tried to start their own commercial timber production. Local conflicts over these issues have increased over the years, and communal–cooperative timber production has not always proved effective, given the lack of state support and the activities of private companies in the region.

The conflicts in the lowland Mixe are in some ways more intense than in the Mixe Alta, given the higher fertility of lands in the region, the still extensive stands of rainforest and deciduous forests, and the attraction these lands offer to cattle ranchers. In addition, the Istmo has been affected since the nineteenth century by foreign and national investment (mostly for timber extraction), and by the completion in 1908 of the transisthmic railway. Since the 1930s *latifundistas* took over commerce, and in many cases local authorities, establishing the town of Matías Romero as the legal and commercial centre for the lowland Mixe and the Chimalapas area. After the transisthmic road, linking the important parts of Coatzacoalcos on the Gulf coast with Salina Cruz on the Oaxacan Pacific coast, had been built in the 1940s, Matías Romero also became the regional centre for the timber trade. Ranching and 'forest latifundismo', closely associated with the Oaxacan governmental bureaucracy established themselves (González Martínez 1986; UCIZONI 1988). Since the early 1970s, pressure on the land of Mixe and Zoque occurred through the construction and resettlement programme for the Cerro de Oro dam and of the Drainage District of Uxpanapa.

With aggressive governmental promotion of 'community forest businesses' in the sierra region as well as in the Mixe and Chimalapas, conflicts over logging and land invasions by settlers and ranchers (including those from Chiapas) moved the indigenous communities of the lowland Mixe and the largely Zoque and Zapotec villages of the Chimalapas region politically closer together. Direct action by Chimalapas communities against illegal loggers and ranchers from surrounding areas, including Chiapas, saw an increase parallel to the rise in timber extracted in the 1970s. Feeding into and in turn gaining support from the popular movements in Oaxaca that managed to dispose of the governor in 1977, these actions, which included roadblocks, 'kidnapping' and destruction of machinery, and public support – even from the 'yellow' national peasant union (CNC) – forced the Agrarian Reform Ministry (SRA) in 1978 to suspend at least the logging licence of one major firm (Sánchez Monroy) and some Chiapanecos. Although logging continued 'discretely', over the following years, some communities began to cut, process and sell wood through their own organizations. Yet in several villages contraband logging by villagers also took place, and internal problems hindered the smooth running of the cooperative business.

Planning experience

Although protests and logging continued, the region was still on the receiving end of another 'rural development' plan. The lessons of the socio-economic and agro-ecological effects of the ill-conceived modernization policies that produced 'developments' like the Benito Juárez dam had apparently not been learned: the 'development pole' of Salina Cruz was to receive further support through the Proyecto Chicapa-Chimalapas. Under this plan, initiated in 1982 by the Ministry of Agriculture (SARH), the Río del Corte, in the centre of the Chimalapas tropical forest region is to be dammed, inundating 20,000 hectares of surrounding lands so that the industrial complexes of Salina Cruz and new irrigation schemes on the Pacific coast may receive further water. The project was planned to be finished by 1992, yet the communities had not even been shown any plans of this (González Martínez 1986). Though construction had not been initiated by 1987, logging had already begun and land had been expropriated.

The project appears to be in jeopardy after events in December 1986, which brought the issue of the Chimalapas to national attention. In July 1986 representatives of the Chimalapas communities had met with the governor to protest against the logging by, amongst others, the brother of the governor of Chiapas, General Absalón Castellanos (*El Diario de Oaxaca* 16.7.1986). No action was taken by the government, so that the new governor, Heladio Ramírez López, faced an inter-state crisis: in December 1986 local activists ('Chimas') detained 10 men whom they accused of being illegal loggers, land invaders, narcotraffickers and cattle thieves. Amongst these 10 were the brother of Absalón Castellanos, his nephew, and 4 gunmen equipped with rifles and shotguns. Their base in the forest and centre of operation was set alight, machinery confiscated or destroyed.

The men were taken to the authorities' offices in Santa María Chimalapas where they were held by the communities as security against demands for compensation for illegal and irrational logging of over 80,000 hectares of forests by outsiders over the last years, for the cessation of further land invasions from Chiapas and elsewhere and for the termination of logging licences (*Noticias* 6.12.1986, 7.12.1986). Governmental representatives, police, helicopters and army units from Chiapas descended on the village, quick action was taken to move ahead with the long-demanded implementation of the presidential decree of 1967 allocating over 460,000 hectares of communal lands. Four days later, with the promise by Ramírez López to meet his Chiapaneco counterpart two weeks later, the 'Chimas' released their hostages.

The governors did meet to make promises and to calm the situation, but the critics of state planning and illegal practices had become alerted. The Alianza Ecologista Nacional denounced the SARH for its indiscriminate granting of licences, especially in this, one of the most important ecological complexes in Mexico, while lacking any coherent reforestation policy (*La Jornada* 8.12.1986).

Support for local activists and indigenous peasants came from the growing urban environmentalist movement in Mexico City. The issue of tropical rainforests and damaging logging had reached a new audience in different socio-economic niches in Mexico, which had become aware of the importance of the ecosystem of the Chimalapas region. It is still one of the few intact ecosystems, spanning ranges between 200–2,300 metres high and constituting the highland basin for the main rivers in the south east (see *Ecología, Política y Cultura*, 3: 1987).[22]

Origins of the UCIZONI

The main problems that led to the organization of Mixe into the UCIZONI were, therefore, land problems and the destruction of forest resources. The emphasis on natural resources originated in the attempts by *caciques* to control communal lands and village production, as well as in the history of logging in the Mixe. But the UCIZONI did not come into being in isolation from activism in villages of different indigenous groups in the region. In 1980 the Organization of Defense of Natural Resources and for the Social Development of the Sierra Juárez (ODRENASIJ) was set up in Ixtlán to struggle against private/state exploitation of the Zapotec and Chinantec forests.

Building on the century-long tradition of Mixe resistance, and faced with an increasing pressure from the effects of the post-war development model, the Committee of Defense and Development of Natural, Human and Cultural Resources of the Mixe region (CODREMI) with its base in Tlahuiltoltepec, was set up in 1980. The expansion of interests, objectives and concern to join communities from all over the Mixe is expressed in the name of this organization. By the mid 1980s the CODREMI had effectively been incorporated into the Assembly of Mixe Authorities (ASAM) from the Mixe Alta. There has recently also occurred a distinct broadening of perspective within the movement – away from the criticized 'ethno-populist' ideology. The violence in the Chimalapas and Mixe Baja region, the increasing confrontations in the teachers' movement in the mid 1980s and the emergence of national peasant and indigenous organizations has provided the ASAM with other sources of potential allies. For this purpose, the ASAM acknowledged the work of the UCIZONI, a potentially competing organization in the Mixe Baja which in recent years has gained regional standing for its defence of communities affected by the confrontation over the

Chimalapas forests, over logging in the Mixe Baja, and over human rights violations.

In the Mixe Baja, attempts to use the indigenous supreme council (*consejo supremo*) to organize independently of the CNC and the influence of the state party, the PRI, in the late 1970s had proven futile, given the influence the government has over the council. Instead, several communities from the area of the central lowland village of Guichicovi and other *municipios* in the Istmo de Tehuantepec began to meet and act jointly over issues of *caciquismo*, land invasions, cattle ranching, lack of services, falling production and above all forest exploitation by private companies and individuals. By 1985 the UCIZONI was beginning to form around the four *municipios* with the participation of some thirty communities from the Istmo.

Echoing the approaches of other activist groups, the UCIZONI believes that self-determination by the communities has to be fought for as much as the land has to be defended. Self-determination here also implies recuperation of and control over lands and natural resources, over local and regional decision-making processes, over naming authorities and carrying out community services, education in their own language, re-valuation of their autochthonous culture, and control over the choice of scientific, technological elements to be adopted for their communities' development (Díaz Gómez 1988). Without wanting to create a state in a state, the Mixe want to have control over their own territory and refute central 'federal' state control over local natural resources. The spurious granting of logging licences, and agrarian and agricultural policies and laws that support private interest to the detriment of the communities are thus more distant but no less important targets of the Mixe activists.

ACTIVISM FOR THE CHIMALAPAS AND SUSTAINABLE DEVELOPMENT

Following the events in December 1986, in 1987 representatives from the Chimalapas, the Sierra Juárez, ecology groups, advisers and government representatives met in Oaxaca to discuss problems of conservation and development of natural resources of the region. The UCIZONI and the meeting participants presented detailed studies of the region, its resources, conflicts and problems. These included agrarian conflicts, road construction, irrational and illegal logging, followed by cattle ranching, hydraulic projects, marijuana

211

cultivation and narcotrafficking as well as the exploitation of animal and plant species for export, increasing violence, forest fires and widespread inadequate control by indigenous communities over their municipal authorities. They pointed out the series of attempts by many communities to start off their own organized and more rational timber businesses, most of which had not come to fruition or were short lived.

The meeting and the UCIZONI document presented a series of recommendations aimed at laying the foundations of a new regional development strategy, opening the way for inter-institutional (state, academic, village authority and environmentalist) cooperation in research and evaluation of existing resources and the initiation of new strategies in resource use and protection. The call for the withdrawal of state forces and for the charges against community members to be dropped and the imprisoned to be released was also made (*Ecología, Política y Cultura* 3: 1987; UCIZONI 1988).

In February 1989 the national federation of environmental organizations, the 'Pact of Ecologist Groups' (PGE), with the support of the communities proposed a 'Programme for the Protection and Integral Development of the Chimalapas, Oaxaca'. Negotiations with the Oaxacan government had resulted in a branch of the decentralized regional planning offices, COPLADE, being set up especially for the Chimalapas region. The aim was for communities, state agencies and the PGE to work together in participatory research on ecological resources and agro-silvicultural production, processing and marketing strategies to be used and controlled by the communities (*La Jornada*, 15.7.1989).

The UCIZONI did not restrict itself to this issue, however. Although it continued to coordinate activities to denounce and pressure agencies like the SARH over their irregularities (for instance, through occupying the offices of the SARH in Tehuantepec in May 1988 (*El Universal*, 12.5.1988)), it also acted as a channel for the denunciation of human rights violations in Mixe communities where cattle ranchers, *caciques* and state forces intensified their aggression by burning forests and crops, beatings, arbitrary detentions and alleged torture. The organization also has projects related to health services, the defence and development of autochthonous cultures, of women's rights and of agricultural production, and provides a legal aid service. It carries out economic studies, arranges for workshops on traditional and non-traditional 'low external input agriculture' (LEIA) technologies and has recently obtained government support for the construction of a coffee *beneficio* and packaging plant, to be administered by the organization (*IWGIA Newsletter* 55–6: 1988; *Etnias*, 1(6): 1989; *La otra cara de México* 8: 1989).

The contacts that the UCIZONI has with NGOs working in the agricultural and forestry sectors, and with grassroots support organizations, and its informal social networks with members of the present regional government have made most of its activities and publicity possible. With the more recent government populist scheme of increasing 'participation', including handing the parastatal Coffee Institute to coffee-producer organizations, the UCIZONI has found another space for activism, one it is exploring openly but with care. The UCIZONI considers the acceptance of government funds, in this case, as a tactical decision necessary to secure income and infrastructure for the community coffee and timber production urgently needed in the region. It refutes the criticism that it is cooperating with the new federal government in a policy of *concertación*, a co-optation measure which the UCIZONI secretary general believes they can avoid (*La otra cara de México* 10: 1989).

The UCIZONI is thus a clear example of how the inevitably interrelated 'problem' areas of race, class and environment can be and are addressed by local indigenous people in an organized way. The advisory role of outside specialists (in legal, agro-forestry and communication spheres, for instance) has been important for its practical use and for the contacts established with national and international organizations and publicity channels. In addition, the strengthened environmental movement and growing, largely urban-based concern over rainforests, as well as the assertiveness of indigenous movements in Oaxaca and the south as a whole have aided in gaining support. Also, the *echevarrista* populism of Ramírez López has opened up certain political spaces and attracted international funding which many groups and movements are at present trying to use to their advantage.

So far the UCIZONI has not relied on explicit alliances with political parties but instead tries to negotiate the maximum possible in terms of resources and concessions from the regional and national governments, while insisting on maintaining its independence in organization and objectives. Whether the organization can grow further, maintain a democratic structure and not be centred around new manipulative indigenous elite groups or individuals, and not fall prey to the *concertación* of the government remains to be seen.

On the other hand, the UCIZONI and the neighbouring ASAM have been active for some years and have, particularly in the case of the UCIZONI, begun to provide concrete projects to be developed by communities from different ethnic groups.

CONCLUSIONS: COOPERATION FOR LOCALLY DETERMINED SUSTAINABLE DEVELOPMENT?

Neither the UCIZONI nor the Mixtepec project is as yet as all-encompassing as its participants would wish, but they show both similarities in perception and the idiosyncrasies of each people and their environment. The experiences of these projects are clearly relevant to the debate over community and agricultural 'development' programmes – whether with or without state intervention. The difference between the approaches discussed here and other committed but 'top-down' projects lies in the fact that indigenous peasant communities and organizations are today more assertive – it is they who choose their 'advisers', demanding clearly the specific inputs they require.

The similarities between these two approaches lie not so much in the cross-ethnic organization as in the shared aims of achieving intra- and inter-community unity, preserving and strengthening local indigenous culture and in emphasizing education. They also have in common concern with the issue of control over communal lands, and the defence of their natural resources against private outside interests. These common concerns point to a three-pronged struggle, focused on ethnicity, ecology and economy.

The diversity of initiatives that have emerged from these communities express not only the variety of local conditions, but also differences in perceptions of the most immediate problems and of how these may be solved. These projects, whether cultural, agricultural or in the defence of human rights and self-determination, make clear that these people do not perceive their crisis to be 'environmental' in the Western sense of soil erosion and deforestation alone. Whereas such ecological factors are clearly important, it is falling production, lack of credit, a disadvantageous marketing system, repression, insecurity of land tenure and threatened cultural environments which are the outstanding concerns. For these autochthonous movements, the isolation of problems and concentration on single solutions is simply not a viable approach.

214

To them, in short, ecological deterioration and falling living standards are but symptoms of a wider crisis.

While not delegating the struggle for land and justice to a secondary plane, the UCIZONI and the project in Mixtepec are particularly concerned with ensuring that the immediate and long-term livelihood needs of the families in the communities are met; these approaches are based on the principles of the 'sustainable participatory development' (Varese and Martin 1986), and refuse the imposition of a 'technology push' or of organizational patterns contrary to local patterns.

The project in Mixtepec is still at a relatively early stage in the elaboration of its programme, although its assertiveness may later lead it to make more extensive contacts with other popular sectors and ethnic groups. In Mixtepec, over time, more people have become involved rather than fewer – a pattern of development which is in marked contrast to state-led programmes. Increasing involvement by indigenous subsistence farmers, particularly from hamlets rather than the richer central village, is an indication that they are slowly evaluating the experience of initial participation. The participation of women and farmers from outlying villages depends on both family and community social dynamics and relations with the central village, as well as on experience with state-led rural projects. The production projects still have to become wholly managed by community members themselves; yet outsiders do not pose a problem as they are accepted and act as appropriate advisers. The management of financial resources and dealings with regional, national and international institutions, however, are handled by project participants themselves.

In the end, indigenous, peasant and environmental organizations in the region are trying to create a space for local approaches to be transformed into practical projects for autochthonous development. They are also learning quickly to hold officialdom to its environmental rhetoric by tabling demands in a language that has suddenly become acceptable to a government trying hard to contain protest over social injustice. Autochthonous approaches offer small but very real instances of people taking their lives into their own hands, beginning to frame their future and challenging a system that has worked to their disadvantage. They demand a culturally sensitive ecodevelopment approach and the right to democratic self-determination.

NOTES

1 Jutta Blauert is Honorary Research Fellow at the Institute of Latin American Studies, London. Her research for this essay is based on material gathered for a doctoral thesis; field-work and research were made possible by a grant from the Economic and Social Research Council (UK), and in the last instance by a research fellowship from the Institute of Latin American Studies. Marta Guidi gathered the information presented on the Mixtec project from material collated since 1985 for an MA thesis and through the experience of working with the Proyecto Mixtepec until July 1990.

2 The 'autochthonous development' referred to, in combining elements of sustainable 'ecodevelopment' and 'ethnodevelopment' (as discussed in the Latin American context by, for instance, Bonfil et al. 1982 and Varese 1983), is taken to imply a development process and strategies conceived by indigenous, and non-indigenous, inhabitants of a region or locality in accordance with their socio-cultural, economic and political structures and requirements. 'Ethnodevelopment' points to the need for development on the basis of the vernacular, be it 'traditional' agricultural knowledge, social organization, production relations or culture. 'Autochthonous development', therefore, has at its centre both the issues of sustainable development and of demands for self-determination by indigenous peoples.

3 The definition of 'sustainable development' relied on in this essay is that based on the rather normative definition by the Brundtland Commission (WCED 1987) and the more people-centred approach by Chambers (1986), based on the concept of 'sustainable livelihood strategies'. This conception of sustainable development, like that of autochthonous development, is most similar to the concept of 'another development' promoted mainly since the mid 1970s by the Dag Hammarskjöld Foundation and theorists/activists like the Chilean Manfred Max-Neef (see Max-Neef 1986; DHF 1975).

4 For a history of the Mexican regional and agricultural planning background see, for example, Hewitt de Alcántara 1976; Gates 1988; Barkin 1977.

5 For a more detailed overview of rural development planning and state intervention in Oaxaca see Segura 1988; Piñón Jiménez 1988. For the big dam projects in the Mazatec/Chinantec regions see Boege 1988; and Barabas and Bartolomé 1973.

6 For a discussion of the wide range of political events and development of popular movements (students, teachers, peasants, indigenous groups, trade unions, etc.) in Oaxaca over the last twenty years see Reina and Sánchez Cortés 1988, particularly volume II; Bustamante et al. 1984; Basañez (ed.) 1987; Yescas Martínez and Zafra 1985; de la Cruz 1986.

7 See Yescas Martínez 1982. The bibliography for the COCEI is extensive; good overviews of its emergence and developments can be obtained from Mejía Piñeros and Sarmiento Silva 1987; Binford 1985; and Martínez López 1985.

8 Significantly, in July 1982 the CNPA changed its slogan 'Hoy luchamos

por la tierra y mañana por el poder' to read 'Hoy luchamos por la tierra y también por el poder.'

9 The best known examples besides the UCIZONI, described below, are: the ODRENASIJ (1980–5) – see Martínez Luna 1982 and 1984; the 'Asamblea de Autoridades Zapotecas y Chinantecas' (AAZC) (1981) – see Parnell 1988 and Basañez (ed.) 1987; the 'Asamblea de Autoridades Mixes' (ASAM) in the highland Mixe (1984 – formerly CODREMI) – see *México Indígena*, 5 (27), 1989, pp. 28–31; the 'Movimiento de Lucha Revolucionaria' (MLR) in the Amuzgo region of the Mixteca (1979) – see Gomezjara 1984; and the 'Movimiento de Unificación y Lucha Triqui' (MULT) (1981) in the Triqui region of the Mixteca – see Durand 1989 and Blauert 1990.

10 An institution in charge of administering the distribution and use of the lands of the municipality – officially communal lands.

11 For a wider discussion of the repercussions of migration in San Juan Mixtepec see Edinger 1985 and Guidi 1988a and 1988b.

12 A group of local, rich traders – mostly *mestizo* – of the municipal central village.

13 The *bienes comunales* continued to be held by the *cacique* faction until the end of 1988. Since then, after a decision by the general assembly of the communities, *bienes comunales* has been controlled by the migrants' association.

14 In the early 1980s, two parastatal agencies, PEMEX and URAMEX, found uranium and oil in municipal lands. Teachers, migrants and authorities, aware of the danger for the environment and the life of local inhabitants, opposed their exploration; the issue, however, is still not resolved. There are also private timber companies logging in forests on the periphery of the municipality, but as yet the villages have not been able to find a legal way to stop these activities.

15 For details regarding the CETAMEX project see Blauert 1990.

16 Since rumour in San Juan Mixtepec is an important element of social control (see Edinger 1985 and Guidi 1988b), the observation of 'benefits' obtained by the first farmers working with the project was rapidly known throughout the municipality.

17 One coordinator, one agronomist, two rural promoters.

18 As this is a community of migrants, where the men remain outside the village for almost the whole year, it is 'normal' in Mixtepec that the women carry out all the work, without seeing their participation in decision-making increase correspondingly; neither has there been a substantial change in their social family status. In this respect, a change in attitude of the new authorities has been relevant, as has the activism of the project in this area (see Guidi 1987, 1990).

19 Given the gradient of most land here, an average plot of 0.25 of a hectare needs three rows of ditches and bunds; for their design and construction at least 8 days of work by 3 or 4 adults is needed.

20 Courses are given by 'Apoyo y Asesoría AC', a Mexican NGO specializing in this work.

21 All of which is coordinated with Mexican NGOs with experience in rural health work.

22 In this way in the last forty years Mexico has lost over 20 million ha in forested area.

REFERENCES

Barabas, Alicia and Bartolomé, M. (1973) *Hydraulic Development and Ethnocide: the Mazatec and Chinantec people of Oaxaca, Mexico*, Copenhagen: IWGIA.

Barkin, David (1977) 'Desarrollo regional y reorganización campesina. La Chontalpa como reflejo del gran problema agropecuario mexicano', *Comercio Exterior*, 27(12): 1408–17.

Basáñez, Miguel (ed.) (1987) *La Composición del Poder; Oaxaca 1968–84*, Mexico City: UNAM/INAPO.

Beaucage, Pierre (1988) 'La condición indígena en México', *Revista Mexicana de Sociología*, 50(1): 191–211.

Benítez Zenteno, Raúl (ed.) (1982) *Sociedad y Política en Oaxaca 1980: 15 estudios de caso*, Oaxaca: IIS-UABJO.

Binford, Leigh (1985) 'Political conflict and land tenure in the Mexican Isthmus of Tehuantepec', *Journal of Latin American Studies*, 17: 179–200.

Blauert, Jutta (1990) 'Autochthonous approaches to rural environmental problems; the Mixteca Alta, Oaxaca, Mexico', Ph.D. thesis, University of London.

Boege, Eckart (1988) *Los Mazatecos Ante la Nación: contradicciones de la identidad étnica en el México actual*, Mexico: Siglo XXI.

Bonfil, G., Ibarra, M. and Varese, S. (1982) *América Latina: etnodesarrollo y etnocidio*, San José: FLACSO.

Bustamante, R., González Pacheco, C. and Lozano, M. (1984) *Oaxaca, una Lucha Reciente: 1960–83*, Nueva Sociología, Mexico (2nd edn).

Chambers, Robert (1986) 'Sustainable livelihoods: an opportunity for the WCED', paper given at the IDS–Sussex University, 13 January, typescript.

de la Cruz, Víctor (1986) 'Reflexiones acerca de los movimientos etno-políticos contemporáneos en Oaxaca', in A. Barabas and M. Bartolomé (eds) *Etnicidad y Pluralismo Cultural: la dinámica étnica en Oaxaca*, Mexico City: INAH, 423–45.

DHF (Dag Hammarskjöld Foundation) (1975) 'El informe Dag Hammarskjöld 1975: ¿Qué hacer?; Otro desarrollo', Development Dialogue, no. 1.

Díaz Gómez, Floriberto (1988) 'Principios comunitarios y derechos indios', paper to the 46th International Congress of Americanists, Amsterdam, July.

Durand Alcántara, Carlos (1989) *La Lucha Campesina en Oaxaca y Guerrero (1978–1987)'*, Mexico: UAChapingo/Costa Amic.

Edinger, Steven (1985) 'Mixtepec. Un pueblo en la sierra', MA thesis, Berkeley: University of California.

Gates, Marylin (1988) 'Codifying marginality: the evolution of Mexican agricultural policy and its impact on the peasantry', *Journal of Latin American Studies*, 20(2): 277–311.

Gomezjara, Francisco (1984) 'Oaxaca: ni "elecciones" ni guerrilla', in R. Bustamante *et al.* (eds) *Oaxaca, una Lucha Reciente: 1960–83*, Mexico City: Nueva Sociología, 271–305 (2nd edn).

González Martínez, Alfonso (1986) *Relación de Santa Maria Chimalapas*, Oaxaca: Casa de la Cultura.

Guidi, Marta (1987) 'Mujer y migración en San Juan Mixtepec', in J. Aranda (ed.) *Las Mujeres en el Campo*, Oaxaca: UABJO.

Guidi, Marta (1988a) 'Estigma y prestigio: la tradición de migrar en San Juan Mixtepec', MA thesis, Social Anthropology, ENAH, Mexico.

Guidi, Marta (1988b) 'Identidad social y destino de los ingresos de los migrantes en San Juan Mixtepec', paper to the 41st International Congress of Americanists, Amsterdam, July, typescript.

Guidi, Marta (1990) *El Saldo de la Migración para las Campesinas Indígenas de San Juan Mixtepec*, Mexico City: Colegio de México (Proyecto de la Mujer).

Hewitt de Alcántara, Cynthia (1976) *Modernizing Mexican Agriculture: socio-economic implications of technological change 1940–1970*, Geneva: UNRISD.

Kearney, Michael (1986) 'Integration of the Mixteca and the Western US–Mexico region via migratory wage labor', in Ina Rosenthal-Urey (ed.) *Regional Impacts of US–Mexican Relations*, San Diego: Center for US–Mexican Studies, University of California, monograph series no. 16, 71–102.

Martínez López, Felipe (1985) *El Crepúsculo del Poder, Juchitán, Oaxaca, 1980–1982*, Oaxaca: IIS-UABJO.

Martínez Luna, Jaime (1982) 'Resistencia comunitaria y cultural popular. El caso de la "Organización en Defensa de los Recursos Naturales y desarrollo social de la Sierra de Juárez AC"', in G. Bonfil *et al.* (eds). *Culturas Populares y Política Cultural*, Mexico City: SEP, 65–78.

Martínez Luna, Jaime (1984) 'Penetración de capital y reproducción comunitaria', MA in Anthropology, University of Veracruz.

Max-Neef, Manfred (1986) 'Human scale economics: the challenge ahead', in P. Ekins (ed.) *The Living Economy*, London: Routledge & Kegan Paul, 45–54.

Mejía Piñeros, M. Consuelo and Sarmiento Silva, S. (1987) 'La lucha indígena: un reto a la ortodoxia', Mexico: Siglo XXI/IIS-UNAM.

Méndez Aquino, Alejandro (1985) *Historia de Tlaxiaco*, published by the author, Mexico City.

Mexico/Oaxaca (1982) *Manual de Estadísticas Básicas del Estado de Oaxaca*, Gobierno del Estado de Oaxaca/SPP (3 vols).

Mexico/Oaxaca (1984) *Programa de Desarrollo Rural Integral de las Mixtecas Oaxaqueñas, 1984–88*, Gobierno Constitucional de Estados Unidos de México y Gobierno Constitucional del Estado de Oaxaca (SPP).

Parnell, Philip (1988) *Escalating Disputes: social participation and change in the Oaxacan highlands*, Tucson: University of Arizona Press.

Piñón Jiménez, Gonzalo (1988) 'Crisis agraria y movimiento campesino (1956–1986)', in Reina and Sánchez Cortés vol. 2, 291–374.

Reina, Leticia and Sánchez-Cortes, J. (eds) (1988) *Historia de la Cuestión Agraria Mexicana: estado de Oaxaca*, Mexico City: Juan Pablos (2 vols – vol. 1: Prehispánico–1924, vol. 2: 1925–1986).

SEDUE (1986) *Informe Sobre el Estado del Medio Ambiente en México*, Mexico City: SEDUE.

Segura, Jaime (1988) 'Los indígenas y los programas de desarrollo agrario (1940–1964)', in Reina and Sánchez-Cortes (eds) vol. 2, 189–290.

SPP/INEGI (1985) *Anuario Estadístico de Oaxaca, 1985*, Mexico City: Gobierno del Estado de Oaxaca/SPP, 3 vols.

UCIZONI (1988) 'Contra los enemigos del bosque', *El Gallo Ilustrado*, 20 March, pp. 2–4.

Varese, Stefano (1983) *Proyectos Étnicos y Proyectos nacionales*, Mexico City: SEP.

Varese, Stefano and Gary Martin (1986) 'El desarrollo participativo sostenido: etnicidad y producción en el trópico húmedo (México y Peru)', paper to the 'Seminario sobre Tecnologías Apropiadas para los Asentamientos Humanos en el Trópico Húmedo' (CEPAL/Secretaría de Planeamiento), Manaos, typescript.

Wall, Miriam (1982) 'Integrated rural development in Mexico – the case study of the Mixteca Alta', M.Phil. thesis, Edinburgh University.

WCED (World Commission on the Environment and Development) (1987) *Our Common Future*, Oxford: Oxford University Press.

Yescas Martínez, Isidoro (1982) 'La Coalición Obrero-Campesino-Estudiantil de Oaxaca; 1972–4', in Benítez Zenteno (ed.), pp. 289–308.

Yescas Martínez, Isidoro and Zafra, G. (1985) *La Insurgencia Magisterial en Oaxaca, 1980*, Oaxaca: IIS-UABJO.

9

RUINING THE COMMONS AND RESPONSES OF THE COMMONERS: COASTAL OVERFISHING AND FISHWORKERS' ACTIONS IN KERALA STATE, INDIA

John Kurien

INTRODUCTION

The last words have yet to be pronounced on the ruin of common property resources and the nature of collective action which is initiated in response to such a situation. Influential opinions on both these issues have, however, greatly conditioned the general thinking on these matters.

As regards the first – the ruin of common property resources – the phrase 'tragedy of the commons', authored by Hardin (1968), has become the stock response when one hears about increasingly numerous examples of the degradation of our planet's common heritage. Hardin pronounced that *whenever* many individuals freely use a common property resource it is doomed to be degraded and will bring ruin to *all*. The emphasis in his article was largely on the *numbers* of 'rational persons' – their increasing population – that take the toll of the commons.

The second issue – collective action vis-a-vis the ruin of a commons – though less well known and discussed, occupies the mind of numerous academics and policymakers (Berkes 1986; Chopra *et al.* 1990; Netting 1981; Oakerson 1988; Ostrom 1989; Runge 1986; Siy 1982). The earliest of these thoughts which tend to dominate current thinking on this issue emanate from Mancur Olson's well known book entitled *The Logic of Collective Action*

(Olson 1965). Olson was of the opinion that the mere presence of a perceived benefit for a group was *not* sufficient to create collective action possibilities to achieve that benefit. He argued emphatically that: 'unless the number of individuals is quite small, or unless there is coercion or some other special device to make individuals act in their common interest, *rational, self-interested individuals will not act to achieve their common or group interests*' (Olson 1965: 2, emphasis in the original).

Hardin's pronouncements and Olson's predilections are not only conditioned by their respective academic penchants, but also very much by the nature of the materialistic and individualistic societies in which their ideas were conceived. These ideas therefore present a very limited perspective of the issues they address, and they are of limited applicability in a cross-cultural context. When commons are seen merely as a source of recreation, and collective action as the privilege of corridor lobbyists, the conclusions of the theoretical work articulated in the 1960s may be valid. However, in the context of the Third World and the vast arenas of interface between common property resources and survival strategies of millions who depend on such resources for a livelihood, there is greater need to delve beyond unidimensional explanations for tragedies and straitjacketed responses to collective action. We need to analyse the numerous, often mutually reinforcing factors that lie behind the ruin of a commons, as well as the plethora of actions – collective and individual – sometimes conflicting and counterproductive, which arise in response to this situation.

In the context of the current enthusiasm for sustainable development and people's participation, the above issues attain a new significance. Common property resources – particularly of the renewable nature – are of prime concern in the sustainable development scenario, and collective action is one important facet which shapes effective people's participation. Sustainable development is premised on a basic notion of intergenerational equity and people's participation postulates a degree of effective collective control in achieving this.

The role of the state is central to the nexus between common property resources and collective action, sustainable development and people's participation. The role of the state in defining the boundaries of common property resources and sustainable development strategies, as well as in prescribing the limits of collective action and people's participation, is well known. And if we do not subscribe to the 'neutrality of the state' theory, we must reckon with the fact that the state's role in delineating the contours of these issues is indeed crucial.

In this chapter we propose to give substance to some of the above thoughts by analysing the economic and ecological crisis resulting from the ruin of a commons – the coastal marine fishing grounds of Kerala State, the south-western maritime province of India – and the responses of the commoners – the traditional, artisanal fisherfolk – to this situation. The attempt will be to highlight this crisis as the result of a *combination* of economic, technological and social factors inherent in a specific context. We will demonstrate that the ensuing detrimental economic and social consequences are by no means equitably distributed. We will also illustrate how the responses at various levels may be collective or individual, and are unlikely to be uniform or necessarily serving to mitigate the crisis. The role of the state and the dilemmas it confronts in striving to cater to the varying interests it serves will also be highlighted. That sustainable development and people's participation are sterile without participatory development and sustainable participation is an important conclusion of the analysis.

The chapter is divided into two main parts. The first deals with the question of the ruin of the coastal commons. Here we begin with a backdrop which very briefly sketches the relevant aspects of the history of the fishery development process in Kerala State. It further enumerates the various factors leading to the overuse of the commons – called overfishing in fishery parlance – provides the available evidence of overfishing, and assesses the varying impact of overfishing on the different interest groups.

The second part deals with the various responses of the commoners and the interface with the state. This is mainly a diachronic narrative of the crucial responses and the various dilemmas faced by the fisherfolk in their pursuit of ensuring a sustainable future for themselves and fishery resources. The manner in which the state attempts to balance the several social forces that place claims on the commons and its produce will also be assessed. Thoughts on ways to resolve the crisis will form a tailpiece.

RUINING THE COMMONS

Backdrop

Fishing, as a subsistence occupation of a caste-bound community, has a long and hoary tradition in India. Traditional marine fishing communities have evolved, over the centuries of learning-through-labour, a keen understanding of the aquatic ecosystem, and have perfected fish harvesting artefacts which were appropriate to that milieu. Their technology was appropriate for fishing merely as a source of meagre livelihood. Such a situation obtained in India until independence in 1947.

Fisheries gained importance with the onset of post-independence economic planning in India. The long coastline and the productive continental shelf gave fisheries the status of a sector capable of accelerating the growth of the rural economy of the country. Accordingly, planned marine fisheries development had the multifaceted objectives of increasing the fish harvest, improving socioeconomic conditions of fisherfolk, augmenting export earnings and generating new employment opportunities. These objectives were to be achieved through initiatives promoted by the state and private efforts.

In order to meet these objectives the 'modernization growth-oriented' model of development, largely premised on the experience of the more developed temperate water maritime countries, was accepted. This approach primarily implied the superimposition of a modern, capital-intensive, specialized technology over the existing traditional base which was largely labour-intensive and of great technical diversity. It assumed that this base was a hindrance to development and had to be either transformed or completely phased out.

By the mid 1960s this development model was introduced in Kerala, in large part because of the rising demand for prawns in the international market. Fisheries development in Kerala State soon became synonymous with increasing prawn harvest and foreign exchange earnings. With the phenomenal rise in the number of small trawlers – introduced initially by the former Indo-Norwegian Project – the prawn harvest and export earnings increased steadily. The earlier caste-bound nature of the fishery sector ceased to be a barrier to entry, and the main investors involved in the new development model were non-fishermen (for details of this see Kurien 1985). For a decade – until the mid 1970s – it was smooth sailing. However, the direction of the tide changed after 1974. The levels of overall fish and prawn harvest began to fall, and by the end of the 1970s the marine fishery sector of the state was heading towards an ecological crisis of overfishing.

The artisanal fisherfolk, who were only peripheral beneficiaries of this modernization model, responded to the crisis at two levels.

The more rapid, widespread and vocal response was in the form of organized protest demanding state regulation of what they perceived as destructive fishing methods. The slower response was in the form of adoption of new technologies for propulsion of their fishing crafts and greater investments in fishing gear in a desperate attempt to enhance their share of falling harvests. They also made more localized attempts to rejuvenate the resource in the coastal commons using their traditional knowledge and reviving age-old practices in this regard.

Overfishing not only implied a fall in the fish harvest but led to a very skewed distribution of the benefits and costs of the fish economy. This in turn came to have larger socio-political implications which plague the state today.

The meaning of overfishing

Overfishing of the near-shore marine waters – the coastal commons – is a problem besetting many developing countries today. The evidence available points to the fact that overfishing has come as a result of many interrelated factors of which the 'common property' nature of these marine waters is but one.

It is customary to distinguish between two types of overfishing: economic and biological. Economic overfishing occurs when marginal costs of an additional unit of fishing effort are higher than marginal revenues. The economy thus experiences loss even though total fish catch may still increase. Biological overfishing occurs when the marginal yield of an additional unit of fishing effort is negative.[1] At such a level of effort the fish population stock is prevented from generating its maximum sustainable yield.[2]

Overfishing thus in a sense heralds a turning point in the dynamics of exploitation of a fishery resource. It is a juncture which if left unattended could spell ruin to much of the fishery resource and to a significant section of those whose lives are dependent on it. In the context of developing countries it would therefore be appropriate to seek policies which avoid problems of excessive effort. This can be achieved through management measures that seek to *maintain* a development process of the fishery which will keep the resource at a high level of productivity by matching fishing effort to the biological and ecological condition of the fish stock. To achieve such a desirable situation presupposes not merely an attack on the *effects* of overfishing, but also a clear understanding of the factors which caused it in the first place.

Factors contributing to overfishing

There are several factors contributing to excessive fishing effort in a fishery. We will restrict our assessment to five major areas which are important in the Kerala context:

1 the open access nature of the fishery;
2 the use of inappropriate technology;
3 the demand-pull factors that create galloping prices;
4 financial subsidies offered by the state which encourage investment; and
5 the pressure of population on the coastal commons.

Open access nature

When traditional technologies and the custom-bound organization of the fish economy predominated, the common property nature of the marine fish resource did not pose a major problem. Technical barriers, such as the need to have fishery-specific skills, and social barriers, such as fishing being the occupation of a lower caste, prevented free entry of capital and persons from outside the traditional fishing communities into the fishery.

The introduction of mechanized boats and the perceived profit opportunities from involvement in activities such as prawn exporting changed this scenario considerably. The vibrant merchant class of Kerala took the first initiatives to break these barriers. They shifted some of their capital from land-based activities to fishing, processing and exporting of prawns. Rapid entry was facilitated by the free access to the sea: mechanized boats could be operated without any form of licence or registration. Entry into the fishery was given greater impetus by the liberal financial assistance of the state (more details about this below). As a result, the post-1966 period witnessed a considerable influx of non-fishermen owners of fishing assets – particularly mechanized trawlers. Between 1966 and 1985 the number of trawlers increased from a couple of hundred to around 2,800.

Use of inappropriate technology

Traditional fishing technologies (nets, tackle and methods of fishing) in general evolved to suit the particular ecological context of the seas and the varying behaviour patterns of the fish. Deserving special mention is the selective nature of fishing nets (a special mesh-size/shape for catching a specific species of fish) and the 'passive' nature of fishing operations (allowing fish to get entangled in the net rather than going in hot pursuit of them or catching them by disturbing their milieu).

As indicated earlier, the 'modernization' phase of fisheries development was premised on the need to introduce fishing crafts, gear and methods which were proven efficient in the temperate water milieu. These tended to be 'active' fishing techniques using single-gear combinations invented for the fishery resources of the temperate waters. Trawling (the method of scraping the sea bottom with a bell-shaped net to catch demersal[3] fish) and purse-seining (the method of quickly encircling whole shoals of pelagic[4] fish) were two such techniques introduced after the decade of the 1960s. Both these techniques were very capital-intensive and initially raised labour productivities. In the short run, unit harvesting costs were low, and given the high prices of certain species of fish (see below), the profits to owners very high. This led to a rapid increase in numbers and the extensive use of these techniques. This contributed very significantly to overfishing by destroying the sea-bottom eco-niche (trawling) and by indiscriminate and non-selective fishing of whole shoals of pelagic fishes (purse-seining).

Booming demand

The introduction of trawlers into Kerala coincided with the rise in demand for prawns in the international market. This was spurred by factors such as the enhanced growth of the US and Japanese economies and also the former's loss of access to supply from China. These demand-pull factors were outside the control of the local economy, and it was difficult to prevent fishery resources from being harvested in response to them.

From a commodity formerly used to provide manure for coconut palms, prawns grew to become the 'pink gold' of marine exports from India. In 1961–2 the beach price of prawns was only Rs 240 per tonne, but by 1971–2 prawn prices reached Rs 1,810 per tonne.

Between then and 1984–5 prices increased nearly seven-fold while the prices of oil sardines and mackerels rose by 184 and 213 per cent respectively. In the case of the domestically consumed fish species – oil sardines and mackerels – there is evidence to show that the increased prices were the result of the inability to enhance the harvests in keeping with the growing demand for fish from the local population (Kurien 1978).

State subsidies

Following the adoption of the 'modernization path' to fisheries development, the state became actively involved in promoting the direction of investments in the sector, instituting many attractive subsidies, investing in capital-intensive and long-gestation infra-structure facilities such as harbours and landing centres, and pro-viding training facilities. In theory, all the 1,200 mechanized boats subsidized by the state between 1961–2 and 1977–8 went to fish-worker cooperatives or genuine groups of fishermen. In practice, however, this seldom happened, as is evident from the evaluation of these cooperatives by a government report which concludes:

> the failure in the operation of the scheme of distribution of mechanized boats were due to the fact that the fishermen cooperatives to whom or through whom the boats were issued were all *benami* (under false name) cooperatives almost with-out any exception. The rich and influential among the fisher-men sponsored and controlled the cooperatives.
>
> (Krishnakumar 1981)

It was this realization which prompted the dropping of a similar scheme drawn up for the sixth five year plan (1980–1 to 1984–5). Subsidy regulations were revised, and from 1985 onwards, state subsidies were enjoyed by genuine fishermen for the first time. The rapid increase in outboard engines in Kerala State, from a handful in 1982 to as many as 8,000 in 1988, is, to a small extent, due to these incentives.

Population pressure on in-shore waters

One characteristic of tropical water fisheries is that overuse of even low productive, passive fishing gear can affect the renewability of stocks (Pauly 1979). The pressure exerted by increasing numbers of fishermen using increasing amounts of fishing equipment within the limited area of the coastal waters has this effect, and population-induced increase of fishing pressure can certainly be viewed as an issue which will now exacerbate the extent of overfishing if present trends continue.

The active fishing population has been increasing at a rate of about 2.3 per cent per annum. In 1961 each fisherman had, on average, 16 hectares of coastal commons to fish. By 1985 the population increased by 65 per cent, reducing the average coastal commons per fisherman to 9 hectares (as against 30 hectares at the all-India level). With the increase in the number of fishermen, total fishing assets also increased. Traditional fishing crafts increased from around 21,000 in 1961 to over 27,000 in 1986. More important are the increases in the quality and the quantity of fishing gear. During the last two decades practically all the fishermen have shifted from using cotton to nylon nets. Though no aggregate estimates are available, evidence from village studies indicate that the quantum of fishing nets and other tackle has increased significantly (Achari 1987a).

The five above-mentioned factors which contribute to over-fishing are complementary and mutually reinforcing. This makes the issue of economic and biological overfishing a very complicated matter to deal with.

The evidence of overfishing

The sea off the south-west coast of India, comprising the maritime states of Goa, Karnataka and Kerala, forms a relatively homogeneous aquatic ecozone, which is the most productive fishing zone in India. Kerala State accounts for just over half of this coastal sea area, which has an estimated maximum sustainable yield (MSY)[5] of 400,000 tonnes (George et al. 1977). The fishery resources in the tropical seas off Kerala State are marked by the multitude of species attaining varying sizes at age of maturity. They are widely dispersed in the coastal commons, and each species is available in relatively small quantities. There are complex prey–predator relationships between them as well as competition for food.[6]

Considerable data are now available to indicate that the above-mentioned factors have, in combination, led to the ecological crisis in the coastal waters of Kerala. The evidence with respect to some parameters is substantial, but is patchy in the case of others. The total picture that emerges, however, points undoubtedly to a scenario of strong tendencies towards overall economic and eco-system overfishing with biological overfishing clearly established in regard to the most valuable species – prawn.

Biological and ecosystem overfishing

Kerala State contributed between 20 and 35 per cent of the total marine fish harvest in India between 1956 and 1985.[7] One can discern two distinct phases in this time span of three decades: a phase of steadily increasing harvests (1956–73); and a phase of stagnating or declining harvests (1973–85). This broad periodization is valid whether one considers the total harvest, the harvest of pelagic and demersal groupings or the major economic species – oil sardines and mackerels, and prawns. This is evident from the growth rates shown in Table 9.1 for the two periods mentioned above.

Table 9.1 Compound growth rates of fish harvest of Kerala State[a]

Species groups	Period I 1956–73	Period II 1973–85
Total marine fish harvest	3.23[b]	−1.79[b]
Total pelagic fish harvest	3.19[b]	−0.18
Total demersal fish harvest	3.52[b]	−4.60[b]
Total oil sardine and mackerel harvest	5.01[b]	0.60
Total prawn harvest	6.21[b]	−8.30[b]

Source: Calculations based on fish harvest data provided by Central Marine Fisheries Research Institute, Cochin.
Notes: [a] Estimated using semi-log function.
[b] Significant at 5 per cent.

To establish that a decline in fish harvests points to biological overfishing conventionally requires that at least two more indicators exhibit a downward trend. These are (1) the catch per unit (fishing) effort (CPUE) and (2) the size of the harvested fish species. Such data are available for penaeid prawns – the most important economic species and the most controversial one in regard to the overfishing debate. In the main prawn landing centre in Kerala (Neendakara) the catch per unit effort (CPUE) declined from 83 kg/hr of fishing effort in 1973 to 20 kg/hr in 1984 (George 1988). An analysis by one of the leading fishery scientists of the country also showed significant decreases in the size of the shrimp caught, causing him to warn of the 'depletionary tendencies noticed in the shrimp fisheries of Kerala and another point of concern from the conservation approach' (George 1988).

Another overall indicator, pointing at least to the possibility of ecosystem overfishing, is the decline in the catches of the demersal species of fish. These bottom-dwelling species are seldom affected by nature-induced changes in their ecosystem. Hence, both increases and declines in their harvests can be attributed to man-induced interventions. Between the years 1971–5 and 1981–5 the harvests of nearly all the important demersal species registered a sharp decline (see Table 9.2). This can largely be attributed to excessive or destructive fishing – particularly the use of trawlers.

Table 9.2 Demersal fish harvest in Kerala (thousand tonnes)

Species	1971–5	1976–80	1981–5	% change from 1971–5 to	
				1976–80	1981–5
Catfish	22	11	10	(50)[a]	(55)
Perches	10	16	7	60	(30)
Sciaenids	10	9	5	(10)	(50)
Leiognathus	11	4	5	(64)	(55)
Prawns	59	41	29	(31)	(51)
Others	36	30	38	(17)	6
Total	148	111	94	(25)	(36)

Source: Paul Babu 1982; Govt of Kerala 1985.
Note: [a] Figures in brackets indicate percentage decline.

Economic overfishing

That economic overfishing had set in by the advent of the 1980s can be gleaned from the evidence of profitability calculations made for the trawler fleet at different points in time. In 1968–9 trawlers in Kerala (above 10 metres length) obtained a 14 per cent return on their investment, after depreciation and interest (Govt of India 1971). By 1978 this figure had dropped to 8.6 per cent (Govt of Kerala 1979), while results of an FAO/UNDP-sponsored study indicated that in 1980–1 trawlers landed significantly fewer fish while incurring larger total costs. This resulted in a negative average rate of return (Kurien and Willmann 1982).

All the above figures are averages, and the profitability range was likely to have been large. Despite 'average losses' it is reckoned that as much as a third of the fleet was operating profitably. This fact, coupled with the fluctuating nature of fortunes from fish harvests, provides a strong incentive for marginal loss makers to continue in the fishery. They pin their hopes on a bumper catch in the near future which could wipe out their accumulated losses.

There is another important reason for the continued expansion of the fleet despite the overall profitability decline indicated by the cost-earnings calculations. Having initially obtained subsidies and long-term loans from the state, the owners of several boats have defaulted in their repayments. In fact, since most of them have appropriate political connections the repayment of loans seems more closely correlated to one's contacts rather than the economics of the operation of one's boats. This makes the *private* return from the boats to the owners still lucrative when calculated on the basis of their *own* investments in it! In March 1986, a provisional estimate of the Government of Kerala assessed the total accumulated arrears on government loan repayments due from mechanized boats (mostly trawlers) to be 30 to 40 per cent of the total costs of these boats. The experience of the commercial banks in this regard is unlikely to have been very different.

An Expert Committee was appointed by the Government of Kerala to study the question of resource depletion and overfishing (see below). This Committee was of the unanimous opinion that the investment in Kerala's coastal waters as of 1985 was far above the desirable optimal levels, estimating the extent of overcapitalization in the fishery to be of the order of Rs 530 million – an amount equal to the total development assistance given by the state to the fisheries sector in Kerala during the three decades of planned development (Achari 1987b). Estimates of excess fishing craft in Kerala are given in Table 9.3.

Table 9.3 Estimates of excess fishing craft in Kerala

Craft type	Existing[a] number	Committee[b] recommendation	Excess Number	%
Trawlers	2,807	1,145	1,662	59
Purse-seiners	54	Nil	54	100
Motorized crafts	6,934	2,690	4,244	61
Non-motorized crafts	20,170	20,000	170	negligible

Source: [a] Department of Fisheries (personal request – Sept. 1986, mechanized gill-net boats not accounted here);
[b] Kalawar *et al.*, 1985.

The economic, ecosystem and biological aspects of overfishing are integrally linked. They reinforce a downward spiral which could in time lead to the complete collapse of the fishery.

Impact of overfishing

The impact of overfishing has dampened the growth of the fisheries sector and widened the gap between it and the rest of the state's economy. The major economic brunt of this was borne by the fishworkers and their families, while the adverse nutritional impact was borne by the local consumers – particularly the poorer among them.

Disparity between sectors

Recent estimates made by the state government indicate that the per capita state domestic product (SDP) is increasing faster than the per capita fishery sector product (FSP). In 1973–4 per capita FSP was 18 per cent lower than per capita SDP, while by 1981 it was 30 per cent lower, and by 1987 the gap had increased to 40 per cent. Though the population growth of the fishing community is higher than the state average, this increasing disparity is primarily due to the slower rate of growth of the fishery sector product. This is due to the change in the composition of fish harvests towards species commanding lower market values following the overfishing of high-value species.

Productivity and incomes of fishermen

The productivity of the working fishermen dropped significantly with overfishing. Incomes, however, did not plunge to abysmal levels because shore prices of fish exhibited considerable increases. The trends in productivity and income were similar for both the workers on the mechanized trawlers and the artisanal fishermen working with their traditional crafts. Taking 1974 as a base we see that productivity and income levels declined across the board. Trawler crews which harvested 10 tonnes of fish in 1974 landed only 7.7 tonnes in 1982, and their real per capita incomes fell by 45 per cent. In the case of the artisanal fishermen the extent of setback was similar. Productivity registered a 50 per cent decline between

Table 9.4 Productivity and income of fishermen in Kerala

| Year | Fishermen on trawlers | | Artisanal fishermen | |
	Productivity (tonnes/yr)	Income[a] (Rs)	Productivity (tonnes/yr)	Income[a] (Rs)
1961	n.a.	n.a.	3.54	330
1965	n.a.	n.a.	3.82	380
1969–70	5.15	790	3.34	630
1974	10.04	2,700	3.20	850
1979–80	7.54	2,630	1.78	540
1982	7.70	1,560	1.62	420

Source: Kurien and Achari 1988.
Note: [a] In per capita terms in 1960–1 prices.

1974 and 1982, and real per capita incomes dropped from Rs 850 to Rs 420 (see Table 9.4).

Increased disparities between workers and owners

Overfishing has not only reduced the income levels of the working fishermen, but it has also increased the level of disparity between them and the non-worker owners of mechanized boats. From a small share of 12 per cent of the total value of output of the sector in 1969, the latter's slice of the fish-pie increased to 27 per cent in the boom period of 1974. Thereafter, with the phase of overfishing setting in, their share increased further, reaching 43 per cent by 1982.

With the increase in the number of mechanized boats between 1969 and 1982 the number of owners has increased. This partly explains the increase in their shares. However, assessments of profitability (mentioned in section above: Govt of India 1971; Govt of Kerala 1979; Kurien and Willmann 1982) indicate that until 1980–1 the *net* returns on investment on mechanized boats on the *average* were positive.

Less fish for the local consumers

Fish was at one time considered to be the poor person's protein in Kerala. No more. Viewed from the perspective of the avid fish-eating population of the state, *more* investments for fisheries development have yielded *less* fish for domestic consumption. The availability and quality of fish sold in the markets have deteriorated, and the retail prices have increased faster than the general cost of other food items (Kurien 1984). There is evidence to indicate that middle- and higher-income households are shifting to more readily available and cheaper sources of protein. The poorer consumers do not exhibit easy changes in diet patterns, and are therefore the ones most affected by this scarcity of fish. Per capita availability of locally consumed fish has decreased from around 19 kilograms in 1971–2 to around 9 kilograms in 1981–2 (Kurien 1985).

RESPONSES OF THE COMMONERS

Having provided a backdrop about the coastal commons and the various mutually reinforcing factors which led to a resource crisis within it, and the impact thereof, we will in this part of the chapter focus attention on the collective macro- and meso-level responses and the individual micro-level actions of the commoners – the fisherfolk – to the situation they confronted. Two factors must be stressed. Firstly, these responses are by no means 'anarchic' or 'autonomous' reactions. They are born out of the evolving socio-economic and techno-ecological forces rooted in the very dynamics of the development of the fish economy of the state. Secondly, it is our opinion that, since the resource crisis itself was brought about by a multiplicity of factors, the responses to it will be equally disparate, and will not always move unidirectionally towards resolving it.

The collective responses were marked by the conscious and participatory efforts of the fisherfolk to influence the state to initiate measures to co-manage the commons in ways which would ensure its sustainability. This was, therefore, essentially a political process through which the fishing community hoped to also achieve steps which would result in greater socio-economic equity. To appreciate the implications of this socio-ecological and political movement, it is necessary to digress a bit to describe the factors which laid the foundation for this new form of 'class' unity of the fisherfolk.

The making of a popular movement

Fisherfolk in Kerala State, as in every other part of India, have been at the margins of society – geographically, economically, socio-culturally and politically.[8] The nature of their occupation, which takes the men out to sea and back to the fringes of the land, thus curtailing social interaction, is one of the predominant reasons for this marginalization. Women in some of the communities are involved in taking fish to the market and they undergo a greater socialization. The effects of this rarely percolate into the male-dominated family life. Added to this, the archaic value systems propagated by organized, male-dominated religion – particularly Roman Catholicism and Islam – curb any tendencies towards forms of organizational unity outside time-honoured, conventional, socio-religious expressions. Economic domination, particularly by merchants and middlemen, often from within the community and having considerable influence over organized religious affairs, curtail any form of new economic formation.

The above factors, working in combination, have given credibility to the perception held by political parties in the state that fishing communities are 'vote banks' to be wooed only at election time. The conservative parties, who generally get the open backing of organized religious forces and the economically powerful, have been able to rest assured of the fishing community votes, come what may. Consequently they have considered it unnecessary and even unwise to work among the actual fishermen – strictly preferring to approach them on all counts through their religious leaders or the influential persons in their communities (generally non-fishermen and often merchants). The more progressive political parties with secular policies and working class concerns have considered traditional artisanal fishworkers an 'unstrategic' group whose votes in any case they could hardly hope to get.

It was in this context, in what seemed like a socially isolated community, that a new genre of social activism began in the mid 1960s. Social activists, predominantly from Christian backgrounds and often even with the support of church-related social action organizations, began systematic work among the fishing communities. These interventions gave equal emphasis to development work and awareness building. They raised the need for more social justice and stressed the importance of popular participation by youth and the labouring sections of the fishing communities in the economic and socio-religious realms. Successful fishworker's cooperatives, women's clubs, youth clubs and a variety of people's organizations were the result. These new ventures at no stage acquired the proportion of a mass movement, but by the mid 1970s they were widely scattered along the coast of Kerala and remained as 'critical and creative irritants' on the periphery of broad traditional socio-religious structures.

Having interacted closely with the economic activities of fishing communities – through efforts at providing credit, introducing intermediate technology, facilitating organized fish marketing and forming credit unions among women fish vendors, among other activities – these social activists came to have an intimate working knowledge of the complexities of fishing, the dynamics of the market and the nature of the socio-economic exploitation experienced by the labouring sections of the community. However, all through their involvements the 'bounty of the sea' was taken for granted. The basic problems of fisherfolk were seen to stem from the 'sharks on land' rather than the lack of fish at sea. It was a problem of getting a fair price for fish – not getting fish.

Activist-researchers in Kerala State had, by the end of 1977, made extensive analyses of the marginalization process of the majority of the labouring fishermen. All the data and other objective facts pointed to the disturbing conclusion that fisheries development policies had become grossly divorced from fisherfolk development priorities. The events in the neighbouring states of Goa and Tamilnadu, where traditional fishermen battled with trawlers that mercilessly rammed into their little crafts and cut their nets, added to this perception. In these states, fishermen, with the active animation of non-party social activists, confronted what they perceived as an infringement of their traditional rights at sea by trawlers owned by non-fishermen. They adopted both militant and non-violent means to do this. The ingress of blatant capitalism into the precincts of traditional fishing via the sea became a reality.

Without much delay, the consequences of the confrontations-at-sea in Goa and Tamilnadu reached Kerala. Its primary effect was to create a new sense of unity among fishermen over the common problem of increasing ingress of trawlers into the coastal waters, and the consequent decline in their catches. By the end of the decade of the 1970s the first steps had been initiated towards the formation of an organization to articulate the fisherfolk's protest and channel the spreading unrest. The vast majority of the artisanal fishermen were either owner-operators of small fishing units, or employed as share workers on them. The economic and social distances between them was not very large. An association which could bring them together to pressure the state 'from below' to protect their liveli-hood was sought. After considerable discussion such an association finally took the form of an independent trade union. This was an anomaly in the political context of Kerala State, where trade unions were nearly always associated with political parties.

The primary initiative in forming the union was taken by the small groups of fisherfolk and social activists (mentioned above), working along the coastline of Kerala. This new organization was called the Kerala Swatantra Malsya Thozhilali Federation (KSMTF: Kerala Independent Fishworkers Federation).[9] While the social activists were predominantly Christians, the fisherfolk who rallied around them were from among the Christian, Hindu and Muslim fishing communities. This situation created concern in the established religious circles – particularly the Catholic Church. They considered such a secular movement, in which clergy and nuns played a vital animation role, to be too 'radical'.

Collective action: macro-responses to safeguard the commons

In 1981 the KSMTF spearheaded the movement of artisanal fisherfolk demanding measures to regulate the anarchic and destructive fishing of trawlers in coastal waters. Their primary demand was for a trawl ban during the monsoon months of June, July and August, arguing that it was during this time that many of the important species of fish spawn in the coastal waters. A monsoon trawl ban was highlighted as an essential management measure to prevent *further* marine resource depletion in the coastal waters. A second demand was for effective enforcement of a trawler-free coastal fishing zone reserved exclusively for artisanal fishermen operating non-mechanized craft and for a total ban of purse-seiners from Kerala's waters. Subsidiary demands for greater social welfare measures were also included.

Fisherfolk predominantly from the southern and central districts of the state joined the KSMTF's multi-strategy struggles. A combination of *nirahara satyagrahas* (fasts), *rasta rokos* (road blocking) and massive processions before the government secretariat in Trivandrum, the capital city, were the tactics used to get public attention and action from the government. The impressive turnout of women of the community in the forefront of the processions and the militant, yet disciplined character of the demonstrations, surprised the press and the police – the latter having always been cautious with fisherfolk whom they had always considered to be 'volatile, unruly and easily provoked'.

The awakening and social upheaval of the coastal areas around issues of occupational concern, and the fact that an independent trade union had championed this cause, caught many political parties in the state on the wrong boat. The 'vote bank' concept was called into question. In a geographically elongated state like Kerala, with numerous coastal constituencies, an unpredictable fisherfolk electorate was a condition no political party could risk – irrespective of the colour of their flag and the content of their policies.

All the major political parties, without exception, created new fisherfolk organizations and joined the fray to be able to claim that they also were part of this historic awakening. The left parties had earlier been involved with fisherfolk, particularly in the northern and central districts of Kerala, but hardly on issues of the nature with which they now found themselves confronted. But these earlier associations paid off and they were also able to muster considerable support from fisherfolk in these areas. This, however, did not mean that they could steal the initiative from the KSMTF.

The government in power at that time was a mixed coalition of political parties – those representing the interests of the Muslim and Christian communities, and also some leftist parties which dominated the front. They were confronted with the dilemma of having to please the agitating fisherfolk and the powerful lobby of trawler and purse-seiner owners and the fish exporters. They enacted the Kerala Marine Fisheries Regulation Act which provided legal backing for the zoning of the coastal waters into areas reserved exclusively for artisanal fishermen using non-mechanized craft, and areas beyond this for trawlers and purse-seiners. The implementation of the Act was delayed and the government pleaded its inability to strictly enforce the law due to the lack of technical and financial resources. However, quick measures were taken to implement the welfare schemes – educational grants for children, accident insurance, more liberal credit, housing loans and the like.

The government also appointed an expert committee to look into what they considered the 'scientific and technological issues and assess the socio-economic consequences of the fishery management demands [particularly the monsoon trawl ban] of the fishermen'. This committee included representatives from the scientific community, the state bureaucracy, the artisanal fishworkers' unions and the trawler/purse-seiner owners. The most reputed of the scientists on the committee failed to appear at meetings on the plea that the fishworkers' demands were 'more political than scientific'. He left it to the bureaucrats to resolve the diametrically opposing positions of the fisherfolk and the trawler owners. Understandably the committee could not arrive at any consensus. Its proceedings were concluded with the fishworkers' organizations' representatives presenting a dissenting note to the chairman. A stalemate prevailed.

During the course of the 1981–2 period of struggle, the church hierarchy became more concerned about the militant and confrontational approach adopted by the KSMTF. A major faction of the clergy which had recently become involved with the fisherfolk were apprehensive about the 'leftist' tendencies which they perceived in the slogans and the songs of the movement. They preferred that the focus be on the fisherfolk as a 'community' rather than fishworkers as a 'class'. Gradually the contradiction between the pro-church faction and the secularists became unmanageable. The conservative Christian faction within the KSMTF broke away to form another union which had the explicit backing of the Catholic church hierarchy and the conservative parties. They called the new union the Akhila Kerala Swatantra Malsya Thozhiali Federation (AKSMTF) – All-Kerala Independent Fishworkers Federation. The priests and nuns who continued to support the KSMTF were censured by the church.

Within a few months of this development, fresh elections were announced in the state since the then ruling alliance broke up. The manner of handling the fishworkers' uprising was one of the significant issues of contention between the alliance members. The ensuing election campaign created a situation where the leadership of the KSMTF was pressed to take a political stand in favour of one of the two electoral 'fronts' which were competing in the election. This created the first major contradiction for this 'independent' movement functioning in a highly charged 'party dominated' political context. The KSMTF leadership was in favour of campaigning openly for the new 'front' led by the left parties because of the favourable stand they had adopted to the 1981 struggles.

After the elections, a new political alliance came to power. It was a conservative alliance with the Congress (I) (the party in power at the federal government of India) as the dominant party. However, the swing in the coastal votes towards the left did not go unnoticed. The ruling party realized the gravity of the situation and its future electoral implications. The importance which they accorded to the turmoil in the fishery sector can be gauged by the fact that for the first time in the history of the state the chief minister held the fisheries portfolio, which was considered until then to be a relatively minor responsibility.

The chief minister was determined to break the power of the KSMTF and the fisherfolk organizations of the left parties. He was able to co-opt AKSMTF and refused to have discussions with the KSMTF. He sought to neutralize the social upheaval by promises of an economic and technological package to be made available directly to individuals and groups of artisanal fishermen at the village level. This was also perceived to be an effective means of softening the militancy of the unions which opposed the government. The government set up a new fisheries cooperative federation to cater exclusively to the needs of the artisanal fishermen, and began to actively promote the emerging trend among them to use outboard motors on their traditional crafts.

Individual responses: enter the outboard motor

While the artisanal fisherfolk struggled at the state level for management and control of access to the coastal zone, at the level of their own fishing operations they began to be more open to new technological innovations. This was most evident in the central maritime districts of the state. The most important of these new innovations was the outboard motor (OBM) which, when used to propel their craft, would reduce the drudgery of their work and provide the flexibility to enable them to fish deeper waters. The new, and more liberal post-1981 import policies of the Government of India (ruled by the Congress (I)) made OBMs from Japan easily available in the market. What started as a cautious experimentation with OBMs by individual fishermen soon acquired the tacit support of the new government in Kerala State.

Initially fishermen using motors were seen to harvest more fish than those who continued to operate the non-motorized craft. Non-owner fishermen exhibited a definite preference for working on motorized units. Observing these changes, the fishermen who could mobilize the financial resources were spurred to opt for motors. However, if the two desirable conditions of (1) reduction in the drudgery of fishing, and (2) fishing in deeper waters were to be *simultaneously* realized with the use of OBMs, then it implied a significant rise in operating costs *without* reduction in the uncertainty of catching fish due to the unfamiliarity with the deeper water fishing grounds. Confronted with this situation fishermen were left with one option: to continue fishing in the overfished coastal waters for longer periods of time with the OBM-provided flexibility of using more active fishing gear – including mini-versions of trawl nets and purse-seine nets.

Between 1982 and 1984 the number of OBMs issued under government subsidy schemes alone reached 1,900. The economic impact of this quiet wave of change in the artisanal fishery sector did not engage the minds of the KSMTF leadership until they were confronted with strong demands from their central zone units for inclusion of 'subsidies and greater quotas of fuel for OBMs' into the list of demands to the government, second only to the demand for the monsoon trawl ban.

This motorization trend, while it surfaced as a contradiction within the ranks of the KSMTF, did not deflect the union from its prime demand for the monsoon trawl ban. Predictions of an eroding base were belied by the overwhelming response to the KSMTF's call, in 1984, for a renewal of its monsoon agitation.

Back to collective action: the historic 1984 struggle

The primary demands of the KSMTF's 1984 agitation were again for a total ban of trawling during the monsoon months and for stricter enforcement of the zoning provisions of the Kerala Marine Fisheries Regulation Act.

By 1984 the KSMTF had taken the form of an umbrella organization with well-knit, cadre-based and regionally decentralized autonomous units. The agitation call led to a total social upheaval of the coastal belt for well over two months. The predominantly non-violent agitation tactics were occasionally marred by violent encounters between irate fishworkers – men and women – and the police. Many fishworkers and union activists were arrested. The agitation caught the attention of the national media. While serious editorials commented on both the social and the ecological aspects of the movement, it was the KSMTF-sponsored indefinite fast-until-death by a Hindu fisherman and a Catholic nun, in the northern coastal city of Calicut, which became the spotlight of media attention. For the press, sensationalizing the fast, and focusing on the role of liberation theology as a motivating force in the movement, overshadowed the fishery issues and the resource management questions sought to be raised by the KSMTF.

The Calicut fast was supplemented by massive processions of

fishworkers and their supporters in key administrative cities and towns to focus attention on their demands. The main highway crossing the length of the state was blocked with canoes at several points on several days. The railway tracks were picketed and the road to the airport in the capital city of Trivandrum blocked. The spontaneity and massive response to the calls for various demonstrations were the hallmark of the agitation. The spirited involvement of the women and young children of the fishing community were crucial in getting the empathy of the public of Kerala for the issues involved. The KSMTF organized fund-raising campaigns among the public, focusing on the theme that fish consumers should support the cause of the artisanal fishworkers as they were the main suppliers of fish for local consumption.

The left parties which had fishworkers' unions conducted 'solidarity demonstrations', expressing their support for the KSMTF's demands, and tried to use their political clout as the opposition in the legislative assembly to demand a settlement of the issues. To their chagrin they realized that they could not muster the support from fisherfolk in the way accomplished by the KSMTF. The KSMTF, still recovering from the impact of the split in 1982, was also not keen to lose its identity by allying too closely with the left party unions. The compromise was to organize high-profile 'joint struggles' – particularly in the form of relay-fasts by the leaders of the various unions in front of the government legislature buildings in Trivandrum.

The government was firm about its stand vis-a-vis the agitation. It was unwilling to negotiate with the leaders and tried its best to break the agitation using strong-arm tactics. It also attempted to wean away sections of the fishworkers through the influences of religious leaders. These attempts met with limited success.

Due to very strong pressure from the trawler-owner lobby and the prawn exporters, the government was unyielding on the major demand of the three-month monsoon ban. Its spokesmen constantly highlighted the phenomenal 'costs' of such a step: a massive fall in the state's foreign exchange earning contribution and unemployment of trawler crew and processing plant workers. The government also warned against 'militant' unionization and tried to placate groups of fishworkers which they perceived to be only 'peripherally associated to the KSMTF' with direct financial assistance in the form of attractive subsidies and soft loans to buy OBMs and new fishing gear.

Seeing a virtual cul de sac to their agitation and the precarious condition of one of their activists who had been fasting for over 20 days, the KSMTF approached the government with a proposal for an experimental monsoon ban on trawling combined with participatory monitoring of the effect of the ban on the fishery resource. The state government did not concede to consider this request. Instead, due to the political pressure from the opposition (left parties mainly) the government conceded to the demand of appointing an unbiased committee of experts to re-examine the management issues being raised and promised to implement the findings of the committee in full. The new three-man committee was composed of an experienced fishery administrator and two leading fishery scientists. It was significant that the trio were from outside Kerala State. Though never stated explicitly, this was to ensure that the socio-economic and political forces in the fish economy of Kerala State would not bias their findings.

They travelled along the length of Kerala's coastline and met with all the sections and groups which had a stake in the fish economy. The committee submitted its findings in mid 1985. It cautioned the government about the impending crisis which could affect the coastal waters if the existing configuration of fishing assets and fishing effort continued to grow in an unregulated fashion (see the Table 9.3 above). They did not approve the need for a monsoon trawling ban but favoured a drastic reduction of the fleet size of the trawlers to half the then current level. They recommended the use of more passive fishing techniques of the type used by artisanal fishermen, were in strong favour of a total ban on purse-seiners, cautioned the government and the artisanal fishermen about the massive motorization drive upon which they had embarked, and highlighted the need for active fishermen's participation in managing the coastal commons.

The main recommendations of this committee remained on paper only, and efforts by the KSMTF in 1985 and 1986 to commit the government to implement them were futile. In 1987 the government was voted out of power. The swing in the coastal votes against it played an important role in the rout. Six years after their first major agitation in 1981, with three changes of government during this period and a continued crisis in the fishery, the fishworker's organizations in general, and the KSMTF in particular, faced an impasse.

Manoeuvres in the impasse

The marine fish harvest in Kerala continued to stagnate after 1985. The average of the harvest for 1986–7 was only about the same as the 1981–5 average (340,000 tonnes), with the important distinction that the artisanal fishermen using OBMs netted the largest share of the total harvest. The limited success achieved in altering the access rights to the coastal commons through their mass actions led fishermen to focus again on their individual responses to the situation by racing forward with motorization and adoption of new fishing gear. However, this period also witnessed some new group initiatives for rejuvenating the coastal commons.

More motorization and new fishing gear

The continued drive towards motorization made the average level of investments by artisanal fishermen in craft, gear and engines soar. In some districts the increase was almost 10–15 fold above the 1980–1 levels. Ownership patterns in the fishery sector changed as well. In some areas the enhanced investment prompted collective ownership and high owner participation in fishing. In other areas the additional capital requirements were initially financed by local fish merchants, who perceived motorization and the new gear as a sure means of control over greater supplies of fish.

Motorization did result in fishing in deeper waters leading to an increase in physical productivity and harvesting of new species. In the central and northern maritime districts, motorization gave a big boost to the use of fine meshed encircling nets called 'ring seines' to harvest pelagic shoaling fish like oil sardines. These ring seines were nothing but smaller versions of the larger destructive purse-seine nets which the traditional fishermen had vehemently opposed.

The increased harvests often resulted in lower beach prices, since the fishworkers had little control over the marketing. This implied that gross earnings did not rise commensurate with productivity. The higher levels of recurring costs (particularly for fuel and repair of the engines), on the other hand, resulted in cash earnings of fishing units being greatly reduced. Despite all this the evidence available suggests that the incomes of the non-owner workers increased in the initial years of motorization. However, this increase was rather short lived since the need for rapid replacement of the imported engines (the economic lifetime of the engines was rated at

2 to 3 years), and their rising costs, drove the owners into debt traps. An immediate response was to alter the share patterns in favour of capital thus depressing the incomes of the workers.

The economic and the ecological impact of motorization and the introduction of the ring seine created new tensions *within* the traditional fishworker groups. In some areas the traditional fishermen – particularly the older among them – opposed the introduction of the ring seines. They argued that such nets would only accentuate the resource crisis in the long run, although the short-run results in terms of increased catches for a few lucky ones could not be denied.

Quite oblivious of the economic, ecological or social implications of the above, the government, through its fisheries development organization, actively promoted the earlier subsidy scheme for the purchase of outboard motors and introduced a new one for ring seines. This trend further accentuated the new tensions within traditional fishworker groups. In some areas traditional fishermen who continued to use conventional fishing gear violently attacked and burnt the ring seines.

Despite these intra-sector problems the KSMTF claimed that the principal problem in the fisheries sector was still the ecosystem damage of the mechanized trawlers. They argued that this was at the root of the anarchic drive for motorization as well as the adoption of nets like ring seines and mini-trawls. The union leaders pointed out that these were only measures taken to beat the trawlers. They were of the opinion that once this principal problem was addressed with appropriate state intervention, the artisanal fishermen would automatically give up their destructive fishing methods.

In some of the KSMTF strong-holds where the trawler menace continued unabated, the newly acquired speed of the fishing craft as a result of motorization emboldened the fishermen to collectively apprehend trawlers and purse-seiners that violated the zoning regulations, thus forcing the government to take action under the KMFR Act. By taking on the self-appointed role of policing their exclusive-use coastal commons, guaranteed under the KMFR Act, they were bringing to bear on the state apparatus the usefulness and the inevitability of having fisherfolk participate more formally in managing the resource. However, neither the government nor the union wanted the entire responsibility for resource management.

Group initiatives at the mezzo-level: people's artificial reefs

A new group response was the revival of an old practice of creating artificial fish sanctuaries or artificial reefs on the sea floor of the coastal waters (see Kurien 1990). These moves were restricted to the Trivandrum District in the southern end of the state where the OBM drive was not coupled with the introduction of the ring seine or the mini-trawl. Reviving the artificial reef idea was also seen by fishworkers as a means of reviving and reinforcing their knowledge of the marine ecosystem. Between 1985 and 1988, the resurgence of the idea gave rise to a rapid spreading of a movement to create people's artificial reefs (PARs) totally funded and erected by the fishermen – at times with the collaboration of social activists and marine scientists sympathetic to the cause. PARs became the symbols of the attempts of the fisherfolk at 'greening their coastal commons'. Constructing PARs also provided an avenue for the creative use of their accumulated, transgenerational knowledge of the aquatic milieu and the behaviour of fish, which had been relegated with the coming of 'efficient' fishing gear which was without the ecological sophistication of traditional fishing methods. PARs also became appropriate physical structures for a fencing of their exclusive fishing zones against the incursion of trawlers. Finally, PARs, being largely group and village initiatives, also provided the 'bridging initiative' between the macro attempts to manage the resource and the need to evolve lasting institutional forms at the community level which could provide the cultural, socio-economic and political empowerment necessary to sustain the participatory ethos of collective action at the macro level.

The localized success of PARs in rejuvenating the coastal commons, and the higher and more valuable fish harvests obtained from around them, has created sufficient interest on the part of the government to consider sponsoring an artificial reef programme. The KSMTF has however opposed such a move for fear that such sponsorship will deprive PAR construction by the fisherfolk of its spontaneity, diversity and autonomy – three ingredients essential for their success and sustainability.

Trawling bans: economics and politics

In 1988 the KSMTF and the other fishworkers' unions threatened to begin their pre-monsoon agitations. Responding to this situation, the government promulgated a partial ban. All the trawler operating centres in the state – except the largest one, Neendakara – were ordered closed for the months of July and August. The reason given for not closing Neendakara was that the heavy concentration of a marine prawn (*p. stylifera*) in the inshore area during these months would perish if not harvested (mainly by the trawlers) resulting in loss of foreign exchange and employment.

The partial ban turned out to be ineffective. It could not prevent trawlers from the other centres operating from out of Neendakara. The boat owners also went to court charging the government with discriminatory treatment of trawlers located in different parts of the state, and got a delayed but favourable order. The artisanal fishworkers' unions were also unhappy with the situation. There were no significant ecological, economic or political gains from this management measure.

The continued conflict between fishermen using traditional fishing crafts and those using trawlers, as well as the fast emerging conflicts between traditional fishermen themselves (over the use of nets like ring seines), prompted the government to seriously re-examine the overall crisis in the fish economy. The government had before it the recommendations of the two earlier committees. Very few of these recommendations had been implemented. It was, however, deemed necessary to constitute a third expert committee to review the situation once again in the light of the recommendations of the earlier committees. This new committee was headed by a reputed marine biologist and secretary to government, and its terms of reference included: a re-examination of the question of the monsoon trawling ban; the effects of the unprecedented increase in the number of outboard engines and their power rating; and also the rapid increase in the use of gear like ring seines by the traditional fishermen. This expert committee submitted its report to the government in June 1989.

The government decided to immediately implement one of the recommendations made by the committee: a *total* monsoon trawling ban from mid July to the end of August. The other recommendations, which included restrictions on the use of ring seines, limitations of the HP rating of OBMs, measures for protection of estuarine areas, and constitution of a study team to monitor the impact of the trawl ban, were kept in abeyance.

The surprise enforcement of the total ban resulted in bloody confrontations between the enforcement police and the boat owners at Neendakara. The boat owners took the matter to both the High Court and the Supreme Court. Both courts were unwilling to issue a stay order to the government's decision. This legal ruling and the unwavering stand of the government on the matter, despite the possible adverse political fallout, ensured that the ban was fully effective.

The ban did result in a considerable loss of employment for the workers in the processing industry. A fair number of the fishermen from the traditional fishing communities who worked as crew on the trawlers found opportunities to go fishing on the motorized boats operated from their home villages. A large number were, however, unemployed.

The total monsoon trawl ban was the most important fishery management decision made by any government in the country since Independence. In October 1989, two months after the ban was lifted, very large pelagic fish landings were reported from all over the state. It would be wrong to attribute this phenomenon *entirely* to the trawling ban, although both the ruling party politicians and the traditional fishworkers' unions have done so. Much of the credit should go to the yet-to-be-well-understood nature-induced changes in the sea – for example, the effect of enhanced rains and the known cyclic fluctuations of pelagic stocks. However, the total ban on trawling, which resulted in the non-disturbance of the aquatic milieu during the monsoon months, has probably contributed significantly to the more pronounced shoreward movement of the pelagic fish shoals in pursuit of food which is found in abundance in the coastal waters areas cooled by the inflow from rivers swollen by the monsoon rains.

The ability of the motorized units – particularly those using ring seines – to harvest whole pelagic shoals also provides an important reason for the increased harvest *given* the favourable nature-induced conditions and the after-effects of the trawl ban mentioned above.

The increased landings depressed the shore prices, and the retail market prices also dropped drastically. Reminiscent of the 1950s, fresh oil sardines were sold as manure for coconut plantations! It is unlikely that this bumper harvest has had a commensurate positive effect on incomes of fishermen. However, it certainly provided a temporary boost to the nutritional status of fish consumers – particularly the poorer among them.

This increased harvest therefore seems to have been brought

about by a strange combination of factors: largely unpredictable nature-induced processes, strong political will leading to firm management measures and the use of ecologically sound over more efficient harvesting technology.

Evaluating the 1989 ban

In early 1990 the government constituted an inter-disciplinary study-team (recommended by the latest expert committee) to embark on a three-year study of the impact of monsoon trawling bans.

The first analysis of the fishery and the socio-economic impact of the 1989 ban was made by this team. The fish harvest in 1989 was reported to be a record 640,000 tonnes – a clear 170,000 tonnes above the 1988 level. However, the additional income effect of this to the fishermen was virtually nil because of the grossly depressed shore prices. There was a drop in the prawn catches of the trawler fleet in 1989, but the larger fraction of this decline was during the *non-monsoon* period! This highlighted the need for both the fisherfolk and the state to shift emphasis from the narrow focus of a monsoon trawl ban to the management and revival of the resources of the coastal commons as a whole.

In May 1990 the annual state convention of the KSMTF met to evaluate its activities and plan its future course of action. The sessions were marked by elation over the federation's decade of struggles which were the prime mover in the government's decision for a total ban of monsoon trawling in 1989. There was considerable soul-searching over the problems being confronted by the members due to their hasty and across-the-board adoption of motorization and new super-efficient fishing gear. The moral basis of demanding a trawl ban when many of them were now using equally destructive fishing gear came in for sharp internal criticism.

The 1990 trawl ban fiasco

The monsoon rains set in by early June 1990 and all eyes in the fish economy were focused on the government. The success of the 1989 ban, and the credit which the government took for implementing this, led the fisherfolk, the trawler owners and the fishery bureaucracy to assume that a trawl ban in the monsoon of 1990 was a foregone conclusion. Only the date of commencement was to be decided.

There were rumours that the newly elected National Front ruling the Government of India, which was facing a tight foreign exchange situation, was exerting pressure on the state government not to enforce a ban due to its impact on net foreign exchange earnings. As though to confirm the rumours, the Marine Products Export Development Authority (MPEDA) – a Government of India organization for promoting marine exports and dominated by private exporters – announced that it was not in favour of management measures like monsoon trawl bans because of the phenomenal loss of foreign exchange which it entailed. They attempted to prevail upon the state government not to impose a trawl ban in 1990, stating that the loss of foreign exchange due to the 1989 ban was about Rs 300 million.

Meanwhile, a reputed research organization in the state, using the data provided by the MPEDA, estimated that the actual loss of foreign exchange was at most Rs 60 million. It was also pointed out that this loss was in large part due to the fall in international prices and *not* primarily due to the loss of production due to the trawl ban.

The KSMTF activists obtained this research document and decided to *gherao* (encircle) the MPEDA headquarters in Cochin, the business capital of Kerala State, to make their protest known over the falsification of data. They argued that, by providing false information, the MPEDA was attempting to indirectly defame the artisanal fishermen's movement as working against the larger interests of the country.

With no announcement of the trawl ban even after mid June 1990, the fisherfolk became restive. The KSMTF sought to pressure the government. Its president commenced a *nirahara sathyagraha* at the gate of the government secretariat in Trivandrum. The fishworkers' unions of the ruling left parties and the conservative opposition parties held separate rallies demanding the ban announcement.

With the pressure on there was hectic activity. The boat owners' association and the representatives of the export processors met the fisheries minister to put forward their positions on certain trawl ban related matters. The boat owners argued that while they could agree to a short trawl ban in the coastal waters, permission to fish in the deeper waters should be guaranteed. In other words, they were seeking the right of 'innocent passage' through the coastal waters even during the time of the ban. The export processors on the other hand were seeking a 'write-off' on the dues they had to pay to a government welfare fund for fishermen.

On 28 June 1990 the government announced the monsoon trawl ban just as the KSMTF was planning to step up its agitation. However, unlike earlier years it did not specify the duration of the ban. This move was viewed with suspicion by the KSMTF. They had got information that some 'deal' had been struck between the boat owners' representatives and a minister in one of the ruling parties who had close links with the marine export lobby.

The government issued two orders to bring the 1990 ban into effect. The first related to the rationale for the monsoon trawl ban which prohibited trawlers from fishing in the coastal commons. The second pertained to the technical specifications required for trawlers which intended to fish in the deeper waters beyond the prohibited coastal commons.[10]

The boat owners' association took the government to court on this second order. They stated that their boats, though not meeting the specifications of the order, could fish for prawns in deeper waters and in fact always did so. By implication they claimed that the second order was a curb on their fundamental rights to pursue employment. The KSMTF joined the litigation by becoming party to the government's stand. After the initial hearings, the judge constituted a legal commission which was empowered to make a random selection of trawlers provided by the boat owners and actually go out to fish in the sea beyond the limits of the coastal commons to ascertain the validity of the boat owners' claims. The court instructed the government, the police and the appropriate fishery technology institute to make the necessary security and technical arrangements to ensure that the commission could undertake its task objectively. The court also permitted a representative of the government and the KSMTF (since they were party to the litigation) to send a representative on the voyage.

The whole process of the experiment was tension-filled. The selected trawlers sailed out to sea. The commission members and the others accredited to the experiment, along with the police, followed on a larger vessel. On crossing the area of the coastal commons (as defined by the KMFR Act) the larger fishery survey vessel was awaiting them. At the appropriate signal the trawlers released their trawl nets and began to fish. On hauling in the net it was found that fish were caught but that there were no prawns. The commission members (the lawyers) gave their report to the judge. After a final hearing the judge pronounced his verdict. He stated that he was sufficiently convinced that the plea of the boat owners was tenable and suggested that the government should reconsider its stand on the second order on the grounds that the technical specifications stipulated by it seemed too stringent.

The prawn exporters chose the trawl ban period to pressure the government over their claims on the payments to the fishermen's welfare fund. The exporters argued that the turnover tax slapped on them was too high, and would even threaten their very survival, on top of the fact that during the trawl ban they were starved of raw material supplies. Moreover, they argued, since no share of the contribution to the funds was being utilized for the welfare of the processing industry workers they saw no valid reason to continue contributing to it. The fisheries minister refused to yield to this pressure since the fishermen's welfare fund was a major source of funds for 'politically important' social security payments to artisanal fishermen promised by his government. The KSMTF and the fishworkers' unions of the left parties openly supported the government on this stand.

Negotiations with the exporters failed, and they reacted by closing down their processing plants. They refused to buy prawns from the artisanal fishermen who were, at the time of the ban, netting bumper harvests of prawns unhampered by the trawler operations.

The economic boycott caused a crash of prawn prices. As with oil sardines in 1989, prawns suddenly became available in the local markets at rock-bottom prices. Urban consumers, who have always had an affinity for prawns, but never the purchasing power to buy it at the export rates, were suddenly confronted with the prospect of a culinary bonanza at affordable prices.

The boycott exposed the total incapability of the government-sponsored fishermen's cooperative federation to intervene in the market. Equally importantly, it made clear to the KSMTF that their decade-old strategy of struggle for conserving resources without a parallel strategy to ensure that their members are assured a fair price for the produce of their labour could be counterproductive. The credibility of the government and the morale of the fisherfolk were at their lowest ebb.

The earlier-mentioned court verdict came at this critical juncture. In less than 24 hours of receiving the news, the government performed what the KSMTF considered to be an act of total betrayal of the artisanal fisherfolk. It withdrew not only the second order *but also* the first order which related to the monsoon ban, although this was neither challenged by the boat owners nor received any adverse remark by the court. In fact, even after the monsoon ban was thus prematurely lifted after a mere 21 days, very few trawlers went fishing. This was partly due to the adverse weather conditions, but more importantly to the fact that a sizeable number of them had pulled up their trawlers onto dry docks for maintenance jobs hardly expecting such a 'bonus' decision favouring the capitalist class from a left government!

The decision of the government caught all the fisherfolk by surprise. The fishermen's union of the leading left party in the government was holding a rally commending the government for taking the bold step to introduce the trawl ban for a second year when they were informed that the ban had been lifted! The official justification given for the lifting of the ban was that an inverse correlation was noted between the number of ban days and the increase in the fish harvest thereafter. In 1988 there was a near two-month (partial) ban and fish harvest for the year was up by over 80,000 tonnes on the 1987 level; in 1989 with a ban of 40 days the fish harvest for the year was 160,000 tonnes above the 1988 level. By this logic the ban of 21 days in 1990 was expected to result in an increase in fish harvest almost equal to the potential yield of the coastal commons itself! Such perverse logic was an insult to the intelligence of the average citizen of the state, and the *real* motives for the lifting of the ban became the subject of debate in the corridors of power along the coast. It became apparent that when the main left party in the government had to choose between remaining in power with the support of its allies and being committed to management measures which would ensure the sustainability of the commons, they chose the former.

TOWARDS RESOLVING THE CRISIS?

It would be a truism to state that the fish economy of Kerala is in the throes of a crisis. From our analysis it is also clear that, in the long run, it is the coastal commons *and* the working fishermen, rather than the capitalists, that will be most affected.

The primary reason for this is that the capitalists can easily move out of the fishery while the fishermen are more or less tied to it, owing to a lack of alternative economic opportunities. For the fishworkers, their future lies in the sea and its common resources. For capitalists, given their short-term perspective, and under the given conditions of investment, the ratio of profits from in-discriminate harvesting of the commons to the profits from regulated and sustainable harvesting are large. For them it actually pays to bring ruin to the commons!

It is such conflicting motivations and actions which provide the basis for the unequal bargaining power of the two classes and the rationale for the state to regulate the coastal marine waters. An action plan to resolve it is indeed the priority of the day. The objective of any programme of action must be two-fold: (1) to revive the sustainability of the coastal commons and (2) ensure that it provides a basis for a decent livelihood and inexpensive food for as large a population as is possible. The achievement of these objectives demands a policy approach in which development and management of the marine resources and the fish economy are seen as two sides of the same coin.

The scale and type of harvesting technology should be in consonance with the known biological and ecological parameters of the resource. Small-scale fishing crafts using multiple sources of energy, selective fishing gear and operations from decentralized centres along the total length of the coastline should be encouraged. Economically efficient but ecologically destructive fishing artefacts should be strictly controlled irrespective of the user.

The ownership of harvesting technology – fishing craft and gear – should be restricted exclusively to those who are willing to fish. An aquarian reform of sorts to ensure this needs to be enacted by the state. Such a community of workers and working owners should be entrusted with the collective rights and responsibility of managing the coastal commons within the jurisdiction of their decentralized operations at the micro and meso levels.

Conscious efforts to enhance the biological productivity of the coastal waters should be given adequate encouragement. Attempts such as the collective creation and establishment of artificial reefs in coastal waters are good examples of this.

Moving to the hitherto unfished deeper waters is an essential step to reduce the pressure on the coastal commons. This is an arena for diverting some of the excess investments presently in the coastal waters. Making fresh investments in the deep sea should be preceded by thorough resource estimation surveys and economic viability studies. These need not be excessively preoccupied with export potentials. Subsidies to those who move out to these waters may be more economically and socially justifiable.

The above options with regard to conserving and enhancing the fishery resource, the choice, ownership and operation of the technology, as well as the social institutions for management of the resource, provide the basic framework for a fresh policy approach. This will be required to pull Kerala's fish economy out of its ecological crisis and provide a sustainable future for the fishery resources in the coastal commons and the commoners.

At the beginning of a new decade the artisanal fisherfolk of Kerala stand poised with a decade of struggle behind them and an uncertain future ahead. In the words of an old fisherman, 'our only hope lies in the sea, for we know that it belongs to the dead, the living and those yet to be born'. This is an important article of faith for all the fisherfolk of Kerala, particularly at this juncture. It is also a pithy understanding of sustainable development.

To give substance to this article of faith calls for genuine, participatory collective and individual responses that can be woven together to form the fabric of a new development process. It also implies a commitment to a programme of action at the macro, meso and micro levels which will ensure sustainable participation to create a development process which is participatory.

NOTES

1 In tropical multispecies fisheries, biological overfishing may occur even though total catch is still increasing because the decline in yield – or complete extinction – of one or several species may be compensated through higher yields of other species.

2 Biologists further distinguish between 'growth overfishing', 'recruitment overfishing' and 'ecosystem overfishing' depending on which is the most important factor preventing full recovery or growth of the stock (Pauly 1979).

3 Species which generally inhabit the bottom of the sea.

4 Species which are predominantly surface dwelling, and whose life cycles are therefore more prone to influences of oceanographic conditions like changes in water temperature, salinity, dissolved oxygen content and so forth.

5 The maximum sustainable yield (MSY) is subject to changes due to biological and ecological factors. Hence, MSY estimated for a year need not be the same for all years. The estimates quoted in the article are taken from George et al. (1977) and are the only available and comprehensive estimates made so far.

6 These characteristics are distinctly different from those of fish resources in temperate waters, where one finds a relatively smaller number of species which grow to larger sizes and each species is available in

teeming millions. The inter-species interactions are also less complex than what obtains in the tropical waters, making it easier to 'target' fishing operations to specific species.

7 Output figures in this and other parts of the chapter (unless otherwise mentioned) are taken from the published data of the Central Marine Fisheries Research Institute. Price data is taken from the Administrative Reports of the Department of Fisheries.

8 In the caste hierarchy of Indian society, fishing communities figured way down the list – though not the lowest. Despite the conversion of many of them to Roman Catholicism and Islam in the fourteenth and fifteenth centuries, their socio-cultural status in society at large has remained low.

9 Between 1979 and 1981, before it came to be called the KSMTF, the movement went through several changes of name, coverage and character which we do not mention here for want of space.

10 This second order gave specifications which very few of the trawlers in the state possessed. The order would thus prevent trawlers used in the coastal commons setting to sea claiming to be only 'steaming through' to deeper waters. As the state's marine regulation enforcement machinery was weak, this was a sure means to ensure foolproof enforcement of the total trawl ban.

REFERENCES

Achari, T.R.T. (1987a) 'The socio-economic impact of motorization of country craft in Purakkad village: a case study', Trivandrum: Fisheries Research Cell.

Achari, T.R.T. (1987b) 'Maldevelopment of a fishery: a case study of Kerala State, India', paper presented at the FAO Indo-Pacific Fishery Commission, Darwin.

Berkes, F. (1986) 'Local-level management and the commons problem: a comparative study of the Turkish coastal fisheries', Marine Policy, 10: 215–29.

Chopra, K., Dekodi, G.K. and Murthy, M.N. (1990) Participatory Development: people and common property resources, New Delhi: Sage Publications.

George, M.J. (1988) 'Study of shrimp trawling in the south west coast of India – particularly Kerala', Trivandrum: Programme for Community Organisation.

George, P.C. et al. (1977) 'Fishery resources of the Indian exclusive economic zone', Souvenir, Cochin: Integrated Fisheries Project.

Govt of India (1971) 'Evaluation of the programme of mechanization of fishing boats', New Delhi: Programme Evaluation Organisation, Planning Commission.

Govt of Kerala (1979) 'Anjengo fisheries development project: an evaluation study', Trivandrum: Kerala State Planning Board.

Govt of Kerala (1985) Economic Review 1985, Trivandrum: Kerala State Planning Board.

Hardin, G. (1968) 'The tragedy of the commons', Science 162: 1243–8.

Kalawar, A.G., Devaraj, M. and Parulekar, A.H. (1985) 'Report of the expert committee on fisheries in Kerala' Bombay: mimeo.

Krishnakumar, S. (1981) 'Strategy and action programme for a massive thrust to fisheries development and fishermen's welfare in Kerala State (1978–83)', Trivandrum: Govt of Kerala.

Kurien, J. (1978) 'Towards an understanding of the fish economy of Kerala State', working paper 68, Trivandrum: Centre for Development Studies.

Kurien, J. (1984) 'Marketing of marine fish in Kerala state: a preliminary study, Trivandrum: Centre for Development Studies.

Kurien, J. (1985) 'Technical assistance projects and socio-economic change – Norwegian intervention in Kerala's fisheries development', *Economic and Political Weekly*, Bombay: 20 (25–6): A70–A88.

Kurien, J. (1990) 'Collective action and common property resource rejuvenation: the case of people's artificial reefs in Kerala state, India', paper presented at the IPFC Symposium on 'Artificial reefs and fish aggregation devices as resource enhancement and fisheries management tools', Colombo.

Kurien, J. and Achari, T.R.T. (1988) 'Fisheries development policies and the fishermen's struggles in Kerala', *Social Action*, New Delhi: 38 (1): 15–36.

Kurien, J. and Willmann, R. (1982) 'Economics of artisanal and mechanized fisheries in Kerala: a study of costs and earnings of fishing units', Madras: FAO/UNDP.

Netting, R. McC. (1981) *Balancing on an Alp*, Cambridge: Cambridge University Press.

Oakerson, R.J. (1988) 'A model for the analysis of common property problems', in National Academy of Sciences, Proceedings of the Conference on Common Property Resource Management, Washington, DC: National Academy Press.

Olson, M. (1965) *The Logic of Collective Action*, Cambridge: Harvard University Press.

Ostrom, E. (1989) 'Governing the commons: the evolution of institutions for collective action', workshop on political theory and policy analysis, Bloomington: Indiana University.

Paul Babu (1982) 'Report of the committee to study the need for conservation of marine fishery resources during certain seasons of the year and allied matters' Trivandrum: Govt of Kerala.

Pauly, D. (1979) *Theory and Management of Tropical Multispecies Stocks: a review with emphasis on the South-East Asian demersal fisheries*, Manila: ICLARM.

Runge, C.F. (1986) 'Common property and collective action in economic development', *World Development*, 14 (5): 623–35.

Siy, Robert J. (1982) *Community Resource Management: lessons from the Zanjera*, Quezon City, Philippines: University of Philippines Press.

10

FROM ENVIRONMENTAL CONFLICTS TO SUSTAINABLE MOUNTAIN TRANSFORMATION: ECOLOGICAL ACTION IN THE GARHWAL HIMALAYA

Jayanta Bandyopadhyay

SUSTAINABILITY, CONSCIOUSNESS AND ACTION AT THE MICRO LEVEL

Most attempts to understand and define 'sustainable development' have been made in an abstract global context. No doubt there are several major issues that have threatened sustainability at the global level. However, equally significant, if not more so, especially in Third World countries, are the region-specific issues of 'sustainability' which concern the survival of large numbers of people. This makes the challenge of defining and practising 'sustainable development' quite varied in diverse social, political and ecological conditions. While issues such as the depletion of the ozone layer must be addressed at the global level, processes of deforestation or soil erosion need to be understood at the local levels. This chapter is, accordingly, an attempt to examine the types of sustainability consciousness which exist in some remote mountain areas of the Indian Himalayas, and which have emerged from local livelihood concerns.

The important global discourse on 'sustainable development' needs to be strengthened and augmented by two types of input. First, the issue of sustainability has to be examined at the regional and local levels. Second, the question of whether 'sustainable development', as understood and debated on global platforms, has any significance for the people in the most remote marginal rural communities must be addressed. If not, what are the concepts of 'sustainability' for the marginalized? Why have the marginalized taken part in so many known and unknown environmental movements? The discourse on sustainable development can gain much if the consciousness about sustainability among the marginal is examined. In the articulation of 'sustainable development', so much attention, in recent times, has been focused on 'sustainability' in the ecological sense that the very important ongoing debate on the desirable nature of 'development' has been largely marginalized. Before one can arrive at a single global model of sustainable development, it is thus necessary to understand what kind of development is to be sustained.

In an attempt to avoid the confusions and ambiguities associated with the value-laden word 'development', in the present analysis it will be replaced by the word 'transformation'.

The process leading to an understanding of 'sustainability' in the context of micro-level transformations is quite different from a theoretical search for the meaning of sustainable development carried out at the global level. Further, it must be largely based on observations of ecological and social processes and their interpretations within a dynamic framework. This means that, in understanding what constitutes sustainable transformation at the micro level, the enquiry should be:

1 based on a concrete understanding of the ecosystemic processes and characteristics in the micro-region;
2 sensitive to and informed by the details of socio-economic transformation that have taken place and are in progress in the ecosystem; and
3 aware of the popular aspirations and conflicts in relation to these transformations and those planned for in the future.

Since sustainability has to be seen in the dynamic context of transformations, in any region-specific exercise attempting to increase an understanding of sustainability the major directions of socio-economic transformation, together with the actors involved, need to be identified. The question of sustainability, whether social or ecological, is rooted in the nature, speed and the extent of these transformations. Environmental conflicts that could be generated by the transformations are often expressions of a collective concern for sustainability. In this chapter, this exercise of looking for concepts of sustainability at the micro level will be taken up in the context of the mountainous region of Garhwal in the Indian state of Uttar Pradesh. In particular, attention will be drawn to the Tehri-Garhwal region, which has witnessed India's two most important environmental conflicts. The first is the internationally renowned forest conflict known as the Chipko movement, and the second is the 12-year-long people's opposition to the construction of a 260-metre dam on the river Bhagirathi at Tehri. Each has made very important contributions to the clarification of 'sustainability' in the management and utilization of two of the most significant common property resources of mountain areas – forests and water. In this chapter the issue of the forest conflicts in the Garhwal Himalaya will be discussed in the context of possible sources of consciousness about ecological sustainability which have emerged in the Chipko movement. This discussion will be followed by an examination of the concern for sustainability in the movement against the Tehri Dam.

SUSTAINABILITY IN THE DYNAMIC CONTEXT OF MOUNTAIN TRANSFORMATION

The flat plains of the globe have generally been areas of intense and concentrated human economic activities and settlements. The mountains, scattered throughout this planet, have been the site of less intense economic activities and have been primarily looked upon, at least by non-residents, as the origin of resources such as water or minerals. At the most, they have been seen as a peaceful refuge and a source of tourism. However, mountain communities have not been a static socio-economic system, but have undergone transformations most suitable to the characteristic conditions of the mountains.

In order to gain a comprehensive understanding of the background of the threats to sustainability in the context of socio-economic and ecological transformations in the mountains it may be advantageous to first examine the typical mountain characteristics. In an initial description, Jodha (1990) identified six types of mountain characteristics, some of which are common for other non-mountain areas, such as deserts. The first four, namely inaccessibility, fragility, marginality and diversity are described as first-order specificities. Natural suitability or niche and human adaptation mechanisms are described as second-order specificities.

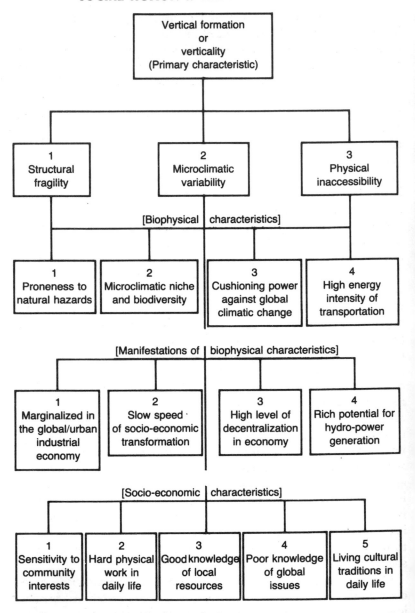

Figure 10.1 Mountain characteristics
Source: Bandyopadhyay 1990

Jodha's description of mountain characteristics puts them in a non-hierarchical but highly interrelated format and is very useful for an understanding of the social and economic dimensions of mountain transformations. Bandyopadhyay (1990a) presents a more complex and hierarchical model of 13 mountain characteristics and identifies only one primary mountain characteristic – that is vertical formation or verticality. Twelve other mountain characteristics are taken to be the result of this primary one. These characteristics are also classified into three categories: bio-physical, socio-economic and cultural, as shown in Figure 10.1. In this way the verticality of the mountains manifests itself not only as a physical barrier, but also leads to a social and cultural discontinuity, in the linkage of human interactive processes, between the plains and the mountains.

Attempts to interpret human economic activities in the mountains have generally been made with the help of altitude zones, a clear manifestation of verticality (Troll 1967). This model worked well in mountain ecosystems under conditions of very limited accessibility and interaction with the plains. When seen in the more recent context of a minimal to strong interaction with the plains and appreciable human accessibility into and within the mountains, the primary characteristic of verticality also manifests itself as an obstacle to easy, quick and cheap physical movements. With the growth of human movements all over the globe and the expansion of human economic activities, such as trade, among different regions, the mountains slowly started to lose their absolute inaccessibility. The isolated mountain societies and the larger settlements in the plains had continuously increasing exchanges and interactions. Since the other two biophysical characteristics of structural fragility and micro-climatic variability are fundamental and largely unchangeable features of the mountains, transformation in the mountains can be interpreted mainly as a factor of changes in accessibility. Allan (1986) has tried to place this decline of inaccessibility (or growth of accessibility) in a model that purports to directly correlate vehicular movement into the mountains and socio-economic transformation.

Once the nature of the transformations in a mountain area is clearly understood, the issues related to sustainability may be identified more easily. Sometimes the transformations initiated by accessibility, governed by the forces of the market, may acquire their own momentum – thus creating conditions for violating the ecological conditions for sustainability. Sometimes enhanced accessibility may initiate a new type of transformation that can lead to sustainability problems in the future, while creating a fair amount of economic growth at the present moment. In those mountain contexts where the use of mechanized transportation, either by surface or by air, has been high, the socio-economic transformations generated by such transportation options become appreciable, at least in pockets. With this introductory background to the rather complex interlinkages between accessibility, transformation and sustainability in the mountains, an analysis of the concrete situation in the Garhwal Himalaya will be made in view of the limited scope of this chapter.

ACCESSIBILITY AND EMERGENCE OF NATURAL RESOURCE CONFLICT IN THE GARHWAL HIMALAYA

The region of Garhwal Himalaya is a small part of the Himalayan arc in the state of Uttar Pradesh in India. It is located about 250 kilometres north east of Delhi and is the source of many important rivers of north India belonging to the Ganga system. The glacial roots of the Yamuna, Bhagirathi, Mandakini and Alaknanda rivers are also the locations of four very important Hindu temples visited by millions of pilgrims every year. In spite of the existence of steep verticality and physical inaccessibility, human movement between the plains of India and these temples has grown continuously. The hard trek to these temples, often lasting for weeks, extends across the ridge of the Greater Himalaya to the temples of Kailash and Mansarovar in the Tibetan Plateau. Other than the pilgrim movement, exchange of goods through trade across the Great Himalayan ridge has also been taking place for more than a thousand years. All transportation, until the beginning of the last century, was based on the use of animal power (mules and yaks), and thus was limited in the quantity and speed of movement.

A new phase in the accessibility of Garhwal, and hence in its economic life, started in 1815 when, following the victory of the British in the Gorkha war, Garhwal was divided into two parts. The western part, comprised of what currently constitutes the Tehri-Garhwal and Uttarkashi districts, came to be identified as the princely state of Tehri. The eastern part, comprised of the present Chamoli and Garhwal districts, came to be known as British Garhwal and became a part of British India. The British annexation led to the beginning of the commercial exploitation of Garhwal's forests, which had, until then, been in use only to satisfy the fuel, fodder and small timber requirements of the local people. In the socio-economic history of Garhwal, this new element of forest exploitation marks an absolutely new phase which requires more detailed description, because in this are rooted the feelings that finally led to the birth of the well-known Chipko movement for forest protection.

ENHANCED ACCESSIBILITY AND EMERGENCE OF FOREST CONFLICTS

During the middle of the nineteenth century, the expanding railway network in the plains of British India required a large number of wooden sleepers. Deodar (*cedrus deodara*) was the ideal timber for this purpose. Taking advantage of the demand for deodar trees, which grow in the upper parts of Garhwal Himalaya, one Mr Wilson, an enterprising Englishman, obtained the lease for exploiting all the forests of Tehri-Garhwal in 1850 from the Tehri state for the paltry annual fee of Rs 400. His economic success was the result of his ability to use the river Bhagirathi and Yamuna to transport timber 200 kilometres downstream to the Indian plains. It was a very important way of reducing inaccessibility almost free of cost.

The huge profits earned by Wilson encouraged the British government of the north-western provinces to enter directly into a contract with the Tehri state in 1865 for long-term felling rights. In the first few years there were systematic surveys and attempts were made to expand forest felling in areas away from the river through the construction of new forest roads. The Conservator of Forests of the NW provinces, Pearson (1869), reported that:

> The Conservator believes that it would not take very much to improve the bed of the river for floating to such an extent that wooden sleepers might be easily floated with safety in the cold weather.... Perhaps the best plan would be to spend a small sum every year in blasting away the worst impediments. ... Two good wire-rope bridges have been substituted for the two most dangerous of the old temporary bridges, and materials are on the ground for the four remaining ones,

which are being put up. When these are completed access to forest will be comparatively safe and easy.

(p. 113)

The construction of forest roads and bridges brought large areas of forest in the catchments of Bhagirathi and Yamuna rivers under commercial exploitation. Increased accessibility led to increased production and export downstream. Guha (1989) reported that between 1869 and 1885, 65 million railway sleepers were exported from the deodar forests of the Yamuna catchment alone. Exploitation was directly related to accessibility and market demand. Seeing the market and profit in forest resource exploitation, the Tehri state in 1895 started its own forest department and declared as reserved or protected most good quality forests and severely limited the people's access to them. The slow but steady result was the growth of people's active opposition to this marginalization and alienation from what they considered to be their common property. The most tragic of these protests, held on 30 May 1930, led to the brutal killing of many protesters in Tilari by the royal army of Tehri. These details are being presented to bring home the point that forest conflicts are not necessarily based on a consciousness of ecological sustainability, as modern environment literature often tends to project. The conflicts over forests resources in Tehri-Garhwal remained for a long time basic economic struggles over access to resources and the right to survival.

THE CHIPKO MOVEMENT AND BEGINNING OF CONSCIOUSNESS ABOUT ECOLOGICAL SUSTAINABILITY

In 1949, the area belonging to the princely state of Tehri joined the Indian Union and became a district in the hill areas of Uttar Pradesh state. In the 1960s, there was a remarkable expansion of road networks in the hill areas of Uttar Pradesh, which drastically reduced the inaccessibility of many corners of Garhwal which had been, until then, considered inaccessible and remote. In this way much larger areas of forests also became directly accessible by road. The forests, normally considered by outside interests as a source of capital to be used in the development activities of a newly independent country, were not the object of ecological sustainability considerations. Thus, official exploitation of forest

266

resources, undertaken by rich and influential contractors from outside the hill areas, was considered beneficial for the country. The poor and marginalized hill people saw in this process a quick draining off of their local resources.

The local people's aspiration to change the forest policy, and to hand over the forest fellings to local cooperatives, led in due course to the Chipko movement, now world famous. The details of the growth and achievements of the Chipko movement have been described in a wide literature (e.g. Shiva and Bandyopadhyay 1986; Guha 1989). The basic conflict between the interests of the local people and the non-local timber contractors is clearly expressed in the poem written during the first Chipko actions in 1973 (Raturi 1989):

> Embrace the trees and
> Save our forests
> Wealth of the Mountain are
> No more for the looters

The poem is clearly an expression of a conflict between local-level survival and growth for the benefit of outsiders. Under conditions of widespread accessibility to the forests of Garhwal, the growth economy of commercial forestry resulted in large-scale conflict with the survival of the agri-pastoral economy of the people of Garhwal. The third economy, the economy of nature (Bandyopadhyay and Shiva 1989) which is directly related to the question of sustainability, was not given any serious consideration even at this stage.

Between 1974 and 1977, however, some important cognitive transformations took place which injected a strong ecological sensitivity into the Chipko movement. One of the central points of enquiry in this chapter is why and how this turn came about. Very few attempts have been made to look into the cognitive linkages behind such a change in order to assess the consciousness of 'sustainability' at the micro level. In an otherwise well researched book, Guha (1989) cuts short his very important observations on this topic. He distinguishes between the 'private' face of Chipko as that of a quintessential peasant movement and its 'public' profile as one of the most celebrated environmental movements in the world. In another attempt, Shiva (1988) tries to make a black and white distinction between the 'feminist' principle of 'conservation' and the 'masculine' principle of 'destruction'. Shiva's analysis seems to be an effort to impose a decadent and outdated Western model of gender conflict on a Gandhian movement characterized by unique gender collaboration. This makes Shiva's work sensational but largely unrealistic when compared to field observations on the same area by Mehta (1990).

THE COGNITIVE ROOTS OF SUSTAINABILITY CONSCIOUSNESS IN CHIPKO

It is widely accepted that the first popular expression of the ecological sustainability of the natural resource base in Garhwal emerged as a slogan in a Chipko meeting in Adwani village in the Henwal Valley of Tehri-Garhwal district. The slogan clearly and concisely articulated the important hydro-ecological processes associated with the mountain forests, and when translated, reads as follows:

What do the forests bear?
Soil, water, and pure air.

The invisible ecological role of forests was, for the first time, openly addressed in the context of the Garhwal region. Any search for the cognitive roots of 'sustainability' consciousness in the Chipko movement should start by looking into the origin of this slogan. Unlike other popular expressions of the movement, it emerged not from the songs written by the celebrated Chipko poet Raturi, but rather from a meeting whose participants were largely village women. Neither the men nor women of Garhwal had stood up against Wilson's earlier tree fellings, which were no less ruthless and took place in a more sensitive catchment. Nor in 1973-4 did the women of Chamoli, who started the Chipko movement and who were no less militant and devoted to the movement, raise such slogans. Obviously, some transformations in the strategy of the women of Garhwal took place during the period from 1973-7. It is important to locate the exact source of this evolution in the ecological understanding of the Chipko movement, which led to its recognition as a leading environmental movement in the world.

The search for the cognitive roots of environmental consciousness in the Chipko movement may thus commence from the distinction made by Guha between the 'private' face and the 'public' profile of Chipko. What transpired between 1973 and 1977 to cause an ecological slogan, one of the most powerful slogans in the world of environmentalism today, to be born in a remote Garhwal village during a meeting of village women? In the following paragraphs an initial explanatory framework within which to answer this question will be presented, although this framework is by no means claimed to be comprehensive.

One of the most important cultural characteristics of mountain societies is their close and detailed knowledge of local resources. Mountain people are obviously much aware of the fragility of their environment and sensitive to any ecological destabilization of it. The close relation with domestic biomass and the knowledge of the hill women about local resources have been very strongly upheld by the leadership of the Chipko movement. Not much attention was initially paid by Chipko to the roles the forests played in soil and water conservation, although these considerations were always a part of the local people's consciousness.

The first direct and formal linkage established between ecological stability and Himalayan deforestation probably came from the report of the Virendra Kumar Committee of the state government, which was entrusted with the task of investigating the possible reasons for the devastating floods in 1970 in the Alaknanda catchment. The committee identified widespread deforestation as a major reason behind the disaster. Almost at the same time, two trend-setting articles by Eckholm (1975, 1976) also appeared. The Chipko leadership was not out of touch with the global currents of thought regarding environmental issues. In particular, Sarala Devi, the mother-figure in the movement and a European disciple of Mahatma Gandhi, had a deep insight both into the local and the global issues and crises emerging from environmental change, and this is embodied in her book *Revive Our Dying Planet* (Sarala Devi 1982). This insight was utilized in her letter to the National Planning Commission of India (Sarala Devi 1978) where she stressed the soil and water conservation properties of the Himalayan forests.

Notwithstanding the fact that Eckholm's claims concerning the direct linkage between Himalayan deforestation and the floods in Bangladesh do not agree with other technical opinions (Hamilton and King 1983), it came as a valuable support to local impressions concerning the hydrological role of the Himalayan forests. This mutual strengthening of the informal local impressions by formal and articulated opinions has proved extremely powerful in various phases of Chipko's resistance to ecological destruction; for example in the case of limestone quarrying in Doon Valley (Bandyopadhyay and Shiva 1989) or more recently in the protest against the Tehri Dam (Bandyopadhyay 1990a and 1990b; Paranjpye 1988). This capacity for 'thinking globally and acting locally' (Rhodes 1990), has been one of the basic strengths of the Chipko movement. The main contribution in this regard came from Sunderlai Bahuguna, the spiritual inspiration of the movement, who played a crucial role in getting the scientific community of India actively involved in supporting Chipko.

It was the result of this capacity of the Chipko movement to bring within its fold the knowledge of a whole spectrum of people, from the common people of the most remote corner of Garhwal to world-renowned foresters like Sir Richard St Barbe-Baker, that its 'public' profile, in due course of time, actually became a major element of the movement. The metamorphosis of Chipko from a peasant movement to a global campaign focusing on the sustainability of forests, on one hand, and sustainability of the agripastoral economy of Garhwal Himalaya, on the other, is real. In India, Chipko is no longer a hill-people's movement against forest felling. It has evolved into a philosophy, a creative extension of Mahatma Gandhi's famous statement on the dichotomy of sustainability and growth: 'The Earth has enough to sustain everyone. But it has got too little to satisfy everyone's greed.'

An analysis of the concepts and expressions of sustainability in the Chipko movement thus brings out the following significant points:

1 The emergence of the Chipko movement as an expression of sustainability consciousness is the result of the leadership's ability to articulate local concerns through global opinions.
2 Popular indigenous consciousness of sustainability in Garhwal, in whatever form it exists, did not find independent utility in spite of a high degree of mobilization on natural resource related conflicts.
3 Indigenous consciousness of sustainability, being more suitable for a local economy with low levels of resource use, becomes useful in handling forest-based conflicts once it also finds exogenous support. This powerful combination was the reason that, in 1981, the government was forced to stop commercial green felling in Garhwal Himalaya for 15 years.

THE ANTI-TEHRI DAM MOVEMENT AND RELATED GROWTH OF SUSTAINABILITY CONSCIOUSNESS

The philosophy of the Chipko movement has expressed itself more concretely in the recent past in a different context, through the emergence of the movement against the proposed high dam at Tehri, the cultural and political nerve centre of Tehri Garhwal. The long drawn out arguments between the officials planning to construct the 260-metre-high dam and the opponents to its construction have thrown up fundamental global issues related to sustainable mountain development, and in particular to the choice of technology for the use of Himalayan water resources. It is to understand, step by step, the growth of sustainability consciousness of the people of Garhwal in particular, and those in the Himalaya in general, that a brief review of the conflict over Tehri Dam will be presented below.

One very important result of the enhanced accessibility of the Himalaya from the plains was the growth of knowledge about its natural resources which were of use to the plains. The water resources of Garhwal Himalaya came as a quick second to the forest resources in drawing the attention of the British rulers, who initiated systematic exploration for these resources. Thus, while the Indian Forest College was started in the foothills at Dehradun, the Thompson Civil Engineering College was established by the British in Roorkee, about 70 kilometres away. Accessibility into the interior Himalaya facilitated the cognitive jump from seeing the Himalayan waters solely in the context of the requirements of the local people, to considering this resource in terms of the needs of the whole river basin. Plans were made for the satisfaction of the water requirements of the far-flung areas in the lower parts of the basin, first by diversion canals and later by the construction of dykes and dams. Like forestry, water resource utilization in the Himalaya seems to have been made the exclusive domain of the interests in the plains. Accordingly, officials from the irrigation or power-generation departments made large-scale plans both to exploit the energy latent in the rivers of the uplands and to use the water for extensive irrigation in the plains. In 1789 the Yaumuna Canal in northern India started a trend of new canals, and in 1930 the Mettur Dam in southern India was the forerunner of an era of large dam construction. The proposed Tehri Dam Project (TDP) at Tehri, for which construction is now underway a kilometre below the confluence of two glacier-fed Himalayan rivers in Garhwal, Bhagirathi and Bhilangana, is among the biggest in the series of dams planned or constructed in the Indian Himalaya.

In spite of their description as 'temples' in official culture, large dams in India were quick to generate popular opposition (D'Monte 1985). The political economy of large dams and the basis of these oppositions have been analysed by Bandyopadhyay (1990a) while Paranjpye (1988) has made an extended benefit–cost analysis of the Tehri Dam Project, identifying the specific reasons for popular protests against this dam. Like the Chipko movement's protests against commercial clear-felling or limestone quarrying in the ecologically sensitive Himalayan catchments, the movement against Tehri Dam has emerged as yet another example of the Chipko movement's ability to use ecological-scientific analysis from diverse sources in strengthening a widespread popular position in favour of using sustainable technologies in the utilization of Himalayan water resources. However, although in many respects the alignment of the social interest groups in the Anti-Tehri Dam movement is similar to that of previous forest movements which took place under the Chipko umbrella, the significance of and the role played by the scientific debate is much bigger in the case of the Tehri Dam. Questions of the meaning of sustainable development and the role which modern ecological science can play in its achievement have come to the fore more strongly than ever before in this mountain region through the Anti-Tehri Dam movement.

The supporters of the dam construction are both locals and non-locals. In spite of the predominantly non-Garhwali labour recruitment in the dam construction so far, the Garhwali middle class holds the fond hope of large-scale employment in the dam at some stage. This group, however, has no idea of the exact number of jobs which will be created by the project: even if all the jobs in the post-construction stage go to the locals, the dam will not make any serious dent on the unemployment levels among the Garhwal youth. There is also a euphoria among the businessmen, traders and large landowners in the urban settlements of Garhwal near Tehri, who believe that the dam construction will mean a bonanza for them because of accelerated growth and escalating land prices in these towns. The non-local groups interested in the dam construction are first, the construction companies and second, the major consumers of electricity – the industrial sector in north India. Politically this is the strongest lobby, which uses the promise of water for Delhi as an attractive wrapping paper to sell the TDP. One also often hears of a strong pro-dam group in the submersion area. It appears that this group is comprised of those who have managed to obtain disproportionately high compensations and have no remaining interest in their homeland.

Among the strongest opponents of the Tehri Dam are the residents of Tehri town and thousands of people of the villages in the submersion area who are dependent on their homeland and do not have the least inkling of where and how they will be relocated. With them are the most articulate, informed and committed anti-dams campaigners – the large contingent of environmentalists in the country. This group collected hundreds of thousands of signatures from all corners of the country in an appeal to the Prime Minister of India to stop this project, whose economic and ecological utility has not been convincingly demonstrated to them, and whose requirement for the large-scale uprooting of about 100,000 people seems excessive. There is another small but important group opposing the dam, which bases its objections on religious concerns about the holy character of the river Bhagirathi, and oppose obstructing its natural flow.

The types of opinions expressed at various stages of the opposition to the construction of the dam have been as diverse as the interest and pressure groups involved in the debate. In 1976 administrative approval to the project was given by the local state government, but the district council of Tehri Garhwal suggested to the state government that the project should be abandoned, in view of the public dissatisfaction over displacement. The natural and instant resistance of the people in the submersion area emerged directly from the threat of displacement and the dismal record of relocation of the past projects. However, the sustained and remarkable efforts of the leadership of the Anti-Tehri Dam Committee (ATDC), in particular of its President Saklani, gradually made the environmental critique and the question of sustainable water technologies the central element in the protest. In this process the opposition to the Tehri Dam brought to the fore a level of scientific understanding which would not have been possible without the movement's initiatives.

It may not be surprising, following the experience of the Chipko movement described above, that this long and important process of environmental evaluation of the Tehri Dam had its origin in impressions which were in fact not technically sound. In 1978, a massive landslide and dam burst occurred on Kanodiagad in the upper catchment area of Bhagirathi. The Anti-Tehri Dam Committee immediately took up this instance as evidence against the stability of the Tehri Dam, and identified deforestation in the upper catchment as the root cause. The Forest Department (FD), however, was not able to identify the disaster as the natural process of mass wasting, characteristic of a growing and unconsolidated mountain. The FD instead claimed that overgrazing was the cause, which in effect implied that the local people were responsible. It is not surprising that the landslide in the Alaknanda catchment generated the ecological thinking in Garhwal's forest movement, while the landslide in the upper Bhagirathi catchment gave an ecological turn to the protest against the Tehri Dam.

The positive outcome of this relatively ill-informed debate was that the Department of Science and Technology of the Government of India established a working group on the 'Environmental impact assessment of the Tehri Dam' in December 1979, of which a courageous environmental crusader, Sunil Roy, was made the Chairman in February 1980. The programmes of the working group under the leadership of Roy set a trend for serious and in-depth impact assessment of Himalayan dams. While Bandyopadhyay (1990a) summarizes important scientific information gaps in the case of the TDP, Dogra (1990) made an attempt to popularize these environmental weaknesses. Slowly, from the large number of articles written in the press, the environmental opposition to the Tehri Dam gained widespread recognition in the country, a development aided by an extremely well-written pamphlet issued by the Anti-Tehri Dam Committee on the occasion of a mass protest rally in Tehri on 31 December 1990 (ATDC 1990).

A very significant aspect of the Tehri Dam was that the construction of the dam was going on during the long years when various studies on its environmental impact were being undertaken. The burden of proof was clearly on the opponents of the dam and not on the planners. Even when doubts about the sustainability of such dams were getting recognition from the scientific committee on large dams in the Department of Environment and Forests, the construction went ahead, creating political pressure against the critics, on the grounds that a poor country like India *cannot* abandon the investments already made, ignoring the argument that a poor country like India should not invest in economically and ecologically unprofitable projects (Bandyopadhyay 1990a). It was ultimately the time tested strength of the non-violent Gandhian strategy of *satyagraha* which forced a temporary halt to the construction. On 25 December 1989, Bahuguna, the philosopher inspiration of the Chipko movement, went to the site of the dam and sat in front of a bulldozer, forcing all construction activities to stand still. In February 1990 the official committee for environmental appraisal of large dams submitted a unanimous recommendation against the TDP. In spite of this, however, the government has not felt pressured to stop further construction of the Tehri Dam. Thus the construction goes on, while environmentalists continue their protest and most people in the submersion area psychologically prepare themselves to leave.

TOWARDS SUSTAINABLE MOUNTAIN DEVELOPMENT: POSSIBLE CONTRIBUTIONS OF THE MOVEMENT

The two cases of environmental movements in the Garhwal Himalaya discussed here, the Chipko movement and the Anti-Tehri Dam movement, are among the most important ones in India. Though the two movements are located in the same area and involved almost the same activist groups, they made very different contributions to the advancement of popular consciousness about sustainable mountain development, and they address sustainability at different planes.

The Chipko movement originated in the conflicts over forest resources. In the context of the agricultural and pastoral economy of the region, the issue of forests has a direct impact on every family living in the region. The protests of Chipko were basically an expression of the threats to the sustainability of the economy of the mountain communities which were generated by commercial forest management. This local issue later took support from global impressions that Himalayan deforestation was causing flood threats to Bangladesh. The Anti-Tehri Dam movement has a much stronger scientific support, to the extent that the official committee on river valley projects expressed an opinion against the construction of the dam.

On the face of it, the forest movement seems to have been a relative success: the demand of the Chipko movement that commercial green felling in the Himalaya be suspended for 15 years was met by the government. The continuation of the construction of the Tehri Dam, on the other hand, appears to be an indicator of the lack of success of the anti-dam movement. A deeper analysis may, however, indicate a greater impact of the Anti-Tehri Dam movement, in terms of the concept of sustainable development, than is evident at first sight.

The positive achievement of the anti-dam movement may be seen in two dimensions. First, unlike the forest movement phase of Chipko, the state has consistently opposed the anti-dam movement. In the case of the demands of the Chipko movement in the forestry sector, the state had accepted demands to end the private contract system for fellings and to a temporary halt of commercial felling. In doing so, the state gained substantial 'green credibility', at least on the surface. The movement against the Tehri Dam was not centred on the issue of the sustainability of a forest culture and an agri-pastoral economy, the flag bearer of tradition. Writings against the dam directly questioned the sustainability of the urban culture and the industrial economy – the flag bearer of modernity. It is here that the anti-dam movement was able to push the state to shed whatever 'green cover' it had, and to stand openly against environmental considerations. It was possible for the state to do so because, in contrast to the forest issue, the dam issue did not unite the rural people in a monolith. The unfortunate households in the submersion area were singled out while the rest of Garhwal was hardly involved. This sends a clear message to all interested in the consciousness of 'sustainability', that when it comes to actual implementations at the micro level, mere green talk on global platforms are of little use.

The second achievement of the anti-dam movement is its ability to raise fundamental questions regarding the nature of development to be pursued. The controversy over the Tehri Dam threw light on the need to rethink the direction of development, and is expressed in the following quote from a pamphlet published by the Uttar Pradesh Sarvodaya Mandal, a Gandhian organization:

The Sarvodaya Mandal of Uttar Pradesh has always opposed the highly centralized development policy because centralized decisions, big-scale technologies, are an indicator of the derecognition of the small common people and destruction of the ecological stability. It is this centralized development which in India has pauperized the majority of the people and encouraged the emergence of violence in all spheres of our life. The proposed Tehri High Dam is only one expression of such faulty development policies. The Sarvodaya Mandal fully supports the popular resistance against the Tehri Dam.

In this respect the anti-dam movement has made a very significant contribution in raising the question of 'what development to sustain and for whom?' In terms of functional reality and of participation of the people, these newly raised questions are of central importance not only to the respective movements but also to the global efforts to define sustainable development. The issues of sustainability in the case of the Tehri Dam are of two types: first is the sustainability of the high-dam technology in the Himalaya; second is the sustainability of the urban-industrial civilization. Both these questions have become live issues for discussion in the villages in Garhwal. The environmental evaluation committee's negative reaction to the dam, and the continued financial support of the state to the project has been a big education to the local people and the environmentalist lobby, demonstrating that mere ceremonial planting of trees does not mean much. If we have to take a serious approach to sustainable development, the negative characteristics of the dominant development concept cannot be ignored. Further, sustainability needs to be defined in terms of continued welfare for all. Accordingly, the nature of socio-economic transformations, unequal access to resources, and localization of growth which results in wider threats to the survival of people in the hinterlands are all factors contributing to the emergence of unsustainability in some form or other.

Thus, before 'sustainable development' can be achieved, a clear understanding of its political economy will be necessary. The roots of unsustainability may be the best point to start the investigation of the political economy of 'sustainable development'. To locate the roots of unsustainability, to listen to the 'South Citizen' as suggested by Nerfin et al. (1975) will be necessary. The search for the consciousness of sustainability in the numerous environmental movements around the world must therefore become an integral part of the global search for 'sustainable development', closely aided by the knowledge of modern ecological sciences.

REFERENCES

Allan, N.J.R. (1986) 'Accessibility and altitude zonation models in mountains', *Mountain Research and Development*, 16 (3): 185–94.

ATDC (Anti-Tehri Dam Committee) (1990) *Why the Opposition to the Tehri Dam*, Tehri, 26 December.

Bandyopadhyay, J. (1990a) 'Tehri Dam: challenge before the NF government', *Economic and Political Weekly*, 25 (5): 243–4.

Bandyopadhyay, J. (1990b) 'Tehri dam: a decision in search of a committee', *Lokyan Bulletin*, 8 (5): 49–53.

Bandyopadhyay, J. and Shiva, V. (1988) 'Political economy of ecology movements', *Economic and Political Weekly*, 23 (24): 1223–32.

Dogra, B. (1990) *A Colossal Risk*, Social Change paper 16, New Delhi.

D'Monte, D. (1985) *Temples or Tombs?*, New Delhi: Centre for Science and Environment.

Eckholm, E. (1975) 'The deterioration of mountain environments', *Science*, 189: 764–70.

Eckholm, E. (1976) *Losing Ground*, New York: Worldwatch Institute.

Guha, Ramachandra (1989) *The Unquiet Woods: ecological change and peasant resistance in the Himalaya*, New Delhi: Oxford University Press.

Hamilton, L.S. and King, P.N. (1983) *Tropical Forested Watersheds*, Boulder: Westview.

Jodha, N.S. (1990) *A Framework for Integrated Mountain Development*, MFS series 1, Kathmandu: ICIMOD.

Mehta, Manjari (1990) 'Cash crops and changing context of women's work and status: a case study from Tehri Garhwal, India', MPE Discussion Paper Series 2, Kathmandu: ICIMOD.

Nerfin, M. *et al.* (1975) 'What now: another development', *Development Dialogue*, 1 and 2.

Paranjpye, V. (1988) *Evaluating the Tehri Dam*, New Delhi: INTACH.

Pearson, G.F. (1869) 'Report on Bhageeruthee Valley forests', NW province government (written from field camp), 7 November.

Raturi, Ghanshyam (1989) *Chipko Songs: Ganga ki Mait Biti*, Uttarkashi: Sarvodaya Kuti.

Rhodes, R.E. (1990) 'Thinking globally, acting locally', paper presented at ICIMOD, Kathmandu, 10–14 September.

Sarala Devi (1978) Letter to the Planning Commission in connection with the formulation of the Sixth Five Year Plan.

Sarala Devi (1982) *Revive Our Dying Planet*, Nainital: Gyanodaya.

Shiva, V. (1988) *Staying Alive: women, ecology and survival in India*, New Delhi: Kali for Women.

Shiva, V. and Bandyopadhyay, J. (1986) 'The evolution, structure and impact of the Chipko movement', *Mountain Research and Development*, 6 (2): 133–42.

Troll, C. (1967) 'Die Klimatische und Vegetationsgeographische gliederung des Himalaya-system', in W. Hellmich (ed.) *Khumbu-Himal*, Heidelburg: Springer-Verlag, pp. 353–88.

Part IV

LESSONS FROM ENVIRONMENTAL PROJECTS

11

ENVIRONMENTAL REHABILITATION IN THE NORTHERN ETHIOPIAN HIGHLANDS: CONSTRAINTS TO PEOPLE'S PARTICIPATION[1]

Michael Ståhl

INTRODUCTION

Many forms of 'participation' are relevant to the topic of environmental sustainability, but this chapter focuses on one form in particular: a programme in which local people are involved in massive campaigns to dig trenches, move stones and plant trees in exchange for food. The emphasis of the chapter is on examining the policies and politics which hamper a more responsible participation by the people of the Ethiopian highlands in the rehabilitation of their farmlands, pastures and hillsides.

The first section of the chapter provides a general description of the character of environmental degradation in the highlands, as well as of the policies designed to combat it. Examples are drawn from Wollo region, which has become notorious for its repeated crop failures and famines. The chapter then discusses the potential for environmental rehabilitation, with a particular concentration on government policies and people's responses to them. The analysis emphasizes these human and institutional factors because degradation and rehabilitation are here considered to be consequences of decision-making by land users who, directly or indirectly, act in response to government policy.[2]

The presentation refers to the period 1985–8. Since that time vast parts of the northern highlands south of Tigray have been overrun by opposition movements and government programmes have ground to a halt.

ENVIRONMENTAL DEGRADATION IN ETHIOPIA

Trends in natural resources use

The thick volcanic soils in the Ethiopian highlands[3] have high inherent fertility and once supported large forests with diversified flora and fauna.[4] The cool climate and ample rainfall attracted early human settlement, and mixed agriculture and stock keeping emerged three thousand years ago. The use of draft animals and ploughs (the ard) was introduced and stable agricultural communities developed. The historical Abyssinian kingdom flourished in these mountains, where peasant production could support a royal bureaucracy and a feudal military aristocracy. Environmental degradation occurred in the vicinity of settlements, but at that time an ample supply of virgin land was available to support the expansion of the growing communities.

Population growth in the twentieth century, enhanced by the partial control of epidemics and by the relatively peaceful period of Haile Selassie's reign after the Second World War, has given a new dimension to the pressure on land. The population growth rate is now estimated to be 2.9 per cent per annum,[5] and the scope for further expansion of cropland is negligible in many parts of the highlands: the land frontier is closed. The impacts of the increased land pressure have been severe. Long fallow periods can no longer be maintained, and arable land comes under continuous cultivation. Soil fertilization through organic manuring has become less frequent because the scarcity of firewood forces people to use cow-dung as household fuel. In the northern parts of the highlands the landscape is generally barren. After centuries of exploitation the forests have been cleared.

A detailed study of the land degradation process in Ethiopia which was carried out in the mid 1980s indicated that erosion is heaviest on sloping agricultural land.[6] Parts of Eritrea and Tigray have already been badly eroded, while the densely populated and intensively cultivated regions of Gondar, Wollo and northern Shoa are now in the frontline of environmental degradation. It is estimated that soil erosion in these areas reduces crop yields by 1 per cent per annum, while biological degradation (a decline in organic matter) may cause a further 1 per cent reduction.[7] Eroded land has progressively shifted to less productive uses as crop yields decrease and the nutritional composition of grasses deteriorate, providing poor grazing for livestock. Eventually, the land produces neither crops nor feed for cattle.

The process of soil erosion in the Ethiopian highlands is exacerbated by both natural factors and agricultural practices. Highly erosive rainstorms hit a landscape that has been stripped bare, with little mitigating cover to protect the soil from the bombardment of the raindrops. Land husbandry practices contribute to the problem. The major crops grown (barley, wheat, teff and sorghum) have small seeds and require a seedbed with fine tilth, which increases the vulnerability of the soil. The consequence is that a high proportion of the rainwater runs off sloping lands either as sheet wash or as torrents, which form deep gullies.

Water erosion is destructive for agriculture in two ways.[8] One is that the uppermost layer of the soil, which is relatively rich in humus, is carried away. The nutrient supply for crops is thus reduced and the soil structure deteriorates. The other negative aspect is that a large part of the water itself runs downstream, ending up in rivers which so far have been underutilized for agriculture in Ethiopia. The availability of water for plant growth is thus reduced. During times when the landscape was generously covered with trees, bushes and grasses, a much higher proportion of the rainwater percolated into the soil and was available in the root zone. High yields in agriculture are, therefore, more dependent today than ever before on a rainfall pattern which guarantees an even distribution of rain over the critical stages of the crop growth cycle. The tendency in the 1980s has been the opposite, however: rainfall has become more erratically distributed, with the result that, even in years where the total annual amount of rain is theoretically sufficient for successful crop growth, poor rainfall distributions often lead to crop failures.

Various scenarios for natural-resource use in the highlands have been developed to predict the future situation in the country in the absence of major changes in agricultural practices, in the rate of population growth or in the rates of resource degradation. One scenario suggests that between 1985 and 2010 soil erosion will cause an increase in the land incapable of supporting agriculture from 2 million hectares to 10 million hectares – an area which covers some 17 per cent of the highlands. Another study suggests that by the same date agricultural production in almost three-quarters of the *awrajas* (administrative districts) will not be enough to provide subsistence for their inhabitants.[9]

A major survey of peasant attitudes to degradation and conservation work has shown that people in the affected areas are well aware of the links between environmental degradation and decreasing agricultural production.[10] In a sample of 2,000 heads of households in 20 peasant associations, the most common factors mentioned as causing soil degradation and erosion were overcultivation, lack of organic fertilizer and the inability to rotate crops. Peasants considered their most serious problem to be the land shortage, which forces them to cultivate their land every year. At the same time, peasants complained about lack of work oxen. A majority of the peasants have only one ox or none, which means that they have to rent oxen from neighbours, and agricultural work proceeds slowly.

When asked why they did not do more to stop soil erosion, the peasants' typical response was that they were too poor and weak, and that they lacked the necessary tools and seeds to do something substantial to arrest erosion. Moreover, the low official market prices paid for the staple crops was mentioned as a disincentive to investing in long-term land improvement. On the other hand, the peasants did not feel that the situation was hopeless. With good leadership in the peasant associations together with technical advice from agricultural experts and assistance, including the provision of improved tools and supplementary food in critical months, they thought it would be possible to reverse the trend towards degradation.

GOVERNMENT POLICY TO COMBAT DEGRADATION

Government programmes to combat degradation include the building of terraces, the closure of hillsides, the planting of trees and the construction of irrigation schemes, as well as the relocation of people on a local level (villagization) and on a regional or national level (resettlement).

Several organizations cooperate in the implementation of these programmes. The World Food Programme (WFP) and the European Community (EC) provide grain and edible oil, while other donors provide hand tools and technical equipment. The peasants, organized in huge working teams, provide the physical labour (such as digging, pitting and planting), while the Ministry of Agriculture (MOA) organizes the work and provides technical supervision as well as training for the peasants. The operations are run as food-for-work programmes.

OVERVIEW OF THE ACTIVITIES

Terraces

Terracing is the recommended treatment for sloping agricultural land, where erosion is most damaging. Stone and soil bunds, which are gradually developed into bench terraces, are constructed, along with grassed waterways and check-dams to lead away surplus rain water. These structures are instrumental in reducing the rate of soil erosion, and in improving the water retention capacity of the soil. In newly terraced areas, springs which have been dried up for years gradually return.

Although few systematic studies of the relation between physical structures and agricultural productivity have been undertaken, the information available indicates that soil bunding has a positive impact on crop yields. The peasants themselves believe, moreover, that on-farm soil conservation activities help stabilize production, mainly through improving the retention of moisture for crops.[11]

Hillside closures

The closure of degraded hillsides is also part of the conservation measures undertaken in the Ethiopian highlands. Under this pro-gramme, the MOA and the local peasant association agree to close certain hillsides from further grazing, cultivation and fuelwood collecting (although cut-and-carry operations to harvest grass are often allowed). Peasant guards enforce the closing. In some cases, trees and grasses are planted, but often the closure is left for natural recolonization. The result is generally impressive. Vegetation returns within a few years and erosion is significantly reduced.

Tree planting

Hundreds of large tree nurseries are run by the government, and thousands of small nurseries are operated by peasant associations in the northern highlands. Together these nurseries have the capacity to raise more than 100 million seedlings annually. Eucalyptus are the most common species planted, followed by cupressus and acacia. Some of the nurseries also produce grass seeds for terrace stabilization and pasture improvement. Trees are planted both on steep slopes for the purpose of soil stabilization, and in conventional woodlots for fuelwood production.

Irrigation

During the 1984–5 famine, the government launched a campaign for the construction of large dams for irrigation purposes, as well as small river diversions for micro-irrigation. The MOA is responsible for small-scale irrigation schemes (below 300 hectares, and in actual fact often only around 10 hectares). These schemes are cheap and can be constructed and managed by peasant associations. The purpose of micro-irrigation is not to create an alternative to rainfed agriculture, but rather to provide facilities for supplementary irrigation. The water sources used are perennial streams and springs. During the 1984 drought, the interest in such micro-irrigation schemes increased, and they mushroomed everywhere. Initially, all the schemes were used for cultivation of cereal food crops, but gradually a diversification towards the production of vegetables for sale in local markets has taken place.

ORGANIZATION OF ACTIVITIES

In the early 1980s, when environmental rehabilitation programmes were first supported by the World Food Programme on a large scale, the emphasis was on stone and earth structures. Since then, the emphasis has gradually shifted to include vegetative measures. Trials are currently being conducted to integrate crop and livestock production with conservation.

Terracing and tree planting are organized in large-scale food-for-work campaigns. Food-for-work is the motivating force in people's participation as workers. With support from WFP, bilateral donors and NGOs, the MOA runs food-for-work activities in nine regions, involving some 800,000 persons and food subsidies of more than 100,000 tonnes per annum. It is the second biggest programme of its kind in the world. In the 1980s, with people mobilized for work through the peasant associations, more than one million kilometres of soil and stone bunds were constructed on agricultural land, close to half a million kilometres of hillside terraces were built, 80,000 hectares of hillsides closed off and 300,000 hectares afforested.

Participating peasants are entitled to a daily ration of wheat (2–3 kilograms) and edible oil (120 grams). In theory the food should be delivered each month. The peasant associations and the MOA are in charge of distribution. Logistical problems with deliveries have been frequent, however, with deliveries sometimes delayed up to six months.

Although food-for-work programmes are sometimes criticized for having a detrimental effect on local food production, in the Ethiopian highlands no such negative impact has been observed. The conservation work takes place during agricultural slack seasons, with the exception of the planting of tree seedlings, which must take place at the onset of rains. Due to the serious food shortages during the 1980s, food deliveries under the food-for-work programme have been a welcome supplement to the locally produced food.[11]

TECHNICAL PROBLEMS

The numerous problems affecting rehabilitation activities have been described and analysed in depth by Yeraswork Admassie and others.[13] A short summary is given here.

Terracing encroaches on the area available for cultivation and provides favourite hiding places for rodents. Terracing may also bring sub-soil to the surface and thereby reduce crop yields.

Most closures lack well-articulated management plans. In the closed hillsides, species with low palatability and nutritional values tend to become dominant, while fodder grasses are further reduced if the hillsides are planted with trees. In addition, the transfer of grazing pressure from the closed areas to the nearby hillsides means that the latter suffer doubly.[14]

The survival rate of tree seedlings is reportedly around 40 per cent. The reasons for this low success rate are that planted species and provenances are not always suitable for local ecological conditions (such as altitude, temperature and soil depth), and that labour for the necessary forest management (thinning, for instance) is insufficient once the plantations have been established. The land assigned to afforestation projects, moreover, is often too marginal to support mature trees.

In 1987, the MOA commissioned an evaluation of the community forestry programme in Ethiopia. Among the socio-economic reasons discovered for the rather weak performance of the programme was that peasants were reluctant to establish woodlots because of their uncertainty concerning the ownership and use rights of these trees and woodlots.[15] The incentive structure will be further discussed below.

Soil bunds on agricultural land, on the other hand, are generally perceived as having positive production benefits. One study reports an increment in production of some 60 per cent on farms where soil bunds were constructed, as compared with nearby farms without bunds.[16]

LACK OF SUSTAINABILITY

Despite all the efforts enumerated above, the major environmental trend in this area continues to be degradation. The conservation programmes affect only a small proportion of the highlands. Away from the major roads, where the great majority of highlanders live, few conservation activities are to be seen. The scope and financing of the programmes are far below needs. However, more important for our discussion is the observation that the conservation programmes in their present form appear to be unsustainable without food-for-work. This is the case in spite of the fact, demonstrated by the above-mentioned survey of peasant attitudes, that people realize the reasons for degradation and are willing to learn and implement improved land reclamation techniques if they receive external help.

It is argued here that a major factor in the lack of independent peasant support for Ethiopia's conservation programmes is the coercive nature of government policies, which have removed people's sense of responsibility for participation in environmental rehabilitation. Before discussing the socio-economic sustainability of the rehabilitation programme, it is necessary to briefly comment on the agricultural policy in a wider context.

THE WIDER CONTEXT OF GOVERNMENT POLICY

In the eroded highlands, government policy has, in addition to the rehabilitation activities reviewed above, emphasized resettlement. Resettlement would, in the official thinking, serve the dual purpose of reducing land pressure in the highlands and mobilizing labour to exploit the vast lowland areas in the south and south west, which traditionally have been sparsely populated.

When drought struck in 1984, the authorities carried out a large-scale resettlement campaign. More than half a million people were resettled in 1984–6, but the campaign has thereafter lost momentum. This is partly due to a lack of resources to administer the resettlement efforts, and partly to the realization among the policy makers that resettlement creates many problems, while it has not yet been proven to solve any. The potential and the constraints, including the health hazards, of the lowlands are yet to be systematically explored.

The situation in the eroded northern highlands is furthermore affected by the general development policy of the government.[17] According to the Constitution of the People's Democratic Revolutionary Republic of Ethiopia, the Workers' Party of Ethiopia (WPE) is the leading force of the state and of the entire society. The WPE systematically promotes a transition to socialism in the countryside. The major ingredients of its programme include state control of the grain trade, the establishment of collective production units (state farms, resettlements and – in the traditional peasant sector – producers' cooperatives) and villagization.[18]

The Agricultural Marketing Corporation (AMC) is the official procurement agency for grains. It establishes prices and requires all peasant households to deliver a grain quota. The producers' cooperatives are agricultural production collectives, in which individual peasant households are encouraged to pool their land, labour and working tools in order to cultivate collectively. Agricultural extension services are wholly concentrated on assistance to these cooperatives. The traditional dispersed settlement pattern is changed by moving all the households which belong to a peasant association into planned, nucleated villages.

The administrative achievements of the state have been substantial in grain marketing, where private traders have been forced either to close business or to function as agents for the AMC. Villagization has also had a substantial impact on rural life, with more than fifteen million people now living in planned villages. The resettlement schemes, although now largely abandoned, were important for a number of years. Collectivization, however, is a disappointment to the officials. Despite continuous agitation, less than 5 per cent of peasant households have joined producers' cooperatives.

The WPE and the state authorities show considerable arrogance in their use of power. They can demand that peasants move into a village within the local peasant association, resettle their households to a foreign environment, or join a producers' cooperative. This arrogance is no doubt responsible for much of the peasants' resistance to government plans. The crux of the matter, however, is that this 'transition to socialism' programme has not proven to be an economic success and offers few opportunities for raising productivity. Surveys indicate that crop yields on cooperative farms are equivalent to or less than those of traditional peasant farms, and that the settlements require continuous subsidies to keep their farms going and to feed the settlers. The AMC's procurement of grain is inefficient and fraught with logistical problems. Villagization has diverted labour and resources – not least trees for new house construction – from production to relocation of homes.

One reason behind the economic failure of these state supported activities is that their focus has been on ideology and organization. Technical and agronomic innovations have received little attention, and no incentives for increased production have been systematically deliberated. A common sight in the agricultural landscape is soil bunds that have been broken down, either by stray livestock or heavy rains, and lie unrepaired. The physical structures need annual maintenance and repair, otherwise they disintegrate, and the transformation of bunds into bench terraces requires continuous build-up and maintenance.

A survey of peasant attitudes to conservation in Wollo clearly demonstrated that, even though peasants appreciate the short-term benefits of increased production from soil bunds, and the potential long-term benefits of trees and revegetated hillsides, they are not prepared to continue conservation work unless supported by food-for-work.[19] Among the reasons mentioned for this attitude in the report on the survey were that peasants had come to expect food payment for maintenance work, and that some conservation measures can decrease agricultural production – terraces, for instance, are sometimes thought to 'steal land'.

The peasants' 'careless' attitude to conservation structures can be interpreted as an alienation from the whole official development strategy. Government policy is designed and implemented in the conventional top-down fashion, with little room for adaption to individual and local conditions. Party and administration officials visit peasant associations and issue instructions and regulations. The often arrogant ways of party and government officials are bound to create frustration, because people are seldom consulted in a genuine manner. They are told what to do, and opposition is considered a counter-revolutionary attitude. Consequently, peasants participate in environmental rehabilitation only when food-for-work is arranged. Their relation to the rehabilitation programme is thus that of paid workers (by the participating peasants the programme is called 'work-for-food') rather than responsible landowners. This clearly shows that the massive popular participation in the official rehabilitation programmes has nothing to do with the empowerment of the people.

Peasants' insecurity also contributes to the land rehabilitation programme's lack of sustainability. The peasants have little room for manoeuvre because, despite the celebrated land reform in Ethiopia, peasants do not own the land they till. The reform abolished absentee landlordism and feudal obligations, and gave use rights to tenants and to former owners who were prepared to cultivate the land with their own labour. But ownership is vested in the state, and the state has shown that it does not respect peasants' use rights when designing grand schemes such as state farms, forest plantations and irrigation projects.

In addition, within the specific context of the rehabilitation policy, the official approach is prescriptive and commandist rather than consultative and supportive of local initiatives. The emphasis is on the number of seedlings planted, rather than on the survival, management and utilization of planted trees; the concern is with how many hectares of hillsides have been closed off, rather than on how to manipulate revegetation. The effect of this commandist approach, combined with the WFP's policy of paying for maintenance work, has been to inculcate attitudes of dependence among the peasant communities when it comes to rehabilitation activities.

One can argue, taking a long-term view, that the main feature of the structural changes in the Ethiopian countryside since the revolution has been to substitute the state for feudal lords as the supreme appropriator of peasant produce and labour. From the point of view of the state, this is a logical step in its attempts at centralization and state building. From the point of view of the peasants, this is something to be avoided, since the state does not use the appropriated resources for improvement of the conditions in the rural areas but rather for consolidating the bureaucracy and waging war on political opponents.

POPULAR PARTICIPATION, THE STATE AND ENVIRONMENTAL REHABILITATION

Popular participation: some remarks

Popular participation is a recurrent theme in discussions on rural development. It emerged as a counter-strategy to efforts to modernize the countryside under large-scale, commercial, mechanized enterprises, which often pushed away the local population. At the most general level the 'participatory approach' calls for rural people to be active participants in efforts to develop their communities in beneficial ways.[20]

The literature on popular participation is voluminous and has produced a multitude of classifications. The concept of participation can be seen as a goal in itself (if the emphasis is on democratic values) or as an instrument in project implementation (if the emphasis is on efficiency). It can focus on contributions to public work schemes, on sensitizing people about development alternatives, on people's involvement in decision-making processes as well as on their control over resources.

UNRISD's research programme on popular participation in the 1970s focused on participation as a question of power, defining the concept as 'the organized efforts to increase control over resources and regulative institutions in given social situations, on the part of groups and movements of those hitherto excluded from such control'.[21] This type of definition of participation is sometimes avoided, since it raises questions about the distribution of power and appropriation of resources in a society. Among donor agencies and national authorities a more favoured approach is to view popular participation as a vehicle for more efficient project implementation at the local level. Nevertheless, in all these conceptualizations, the state, in one form or another, whether the central government, the ministry of agriculture, or the district commissioner, is in opposition to the people. These conceptualizations are reductionist to the extent that they ignore social differentiation.

Since environmental degradation and rehabilitation issues have emerged in the international development debate, the issue of the relationship between popular participation and environment has come to the fore. This relationship is complicated and varies according to such factors as ecological zones, degree of demographic pressure, the options for off-farm employment and land tenure systems.

The situation prevailing in the Ethiopian highlands is characterized by high population pressure, almost total lack of non-agricultural employment, widely shared poverty, technological stagnation and few possibilities for migration (there are strict regulations limiting migration to other peasant associations and to towns). Under these circumstances traditional land husbandry practices, which under former conditions were sustainable, now facilitate degradation and eat into the natural resource base.

In such a situation *the destruction of environment creates value.*[22]

The challenge now is to reverse the situation in ways which will enable peasants to gain profits out of land husbandry methods that rehabilitate the environment. UNRISD's research programme on sustainable development through people's participation in resource management states the problem thus:

> The environmental problem must now be defined not just in terms of the defense of the environment against human use, but how *natural resources can best be managed and exploited creatively for people's benefit*, to optimize their usefulness to the present generation and to maintain and enhance their ability to sustain future ones (emphasis added).[23]

The Ethiopian peasantry is today too exhausted from repeated droughts and civil strife to embark on a transition to sustainable resource use within their local communities. It is obvious that the state and the international community must intervene constructively in this area. As a prelude to our concluding discussion, it is necessary to briefly characterize the Ethiopian state and its capacity for development.

THE STATE

Although Ethiopia has a long history and political tradition, the attempts to build a centralized state are recent. Until the late nineteenth century Ethiopia was known to most foreigners as 'Abyssinia', a region which roughly corresponded to the northern and central highlands in contemporary Ethiopia. This territory was loosely held together by a common culture, of which Monophysite Christianity and the royal tradition of the Solomonic dynasty were the outstanding characteristics. The social structure in these highlands had clearly feudal traits. At the bottom of society were the peasants, who were obliged to render personal services and agricultural produce to their superiors and to follow them in war.

Abyssinia expanded and contracted from time to time. When strong kings ruled, the central administration was enlarged and the power of local lords restricted. At other times, real power was in the hands of regionally based feudal lords. Political and military alliances shifted constantly and warfare was common.[24]

In the latter half of the nineteenth century, centralized political control grew, in part as a response to European imperialism. Eventually Menilek, the king of Shoa, was crowned emperor and embarked on a series of combined diplomatic and military campaigns which gave him hegemony over historical Abyssinia and control over vast tracts of land to the south of Addis Ababa. In this way, the modern state of Ethiopia was created around the turn of the century. To the north, Ethiopia clashed with Italian colonialism. The coastal strip and the Tigray highlands down to the river Mareb, which Italy carved out, became Eritrea.

The reigns of Menilek and his successor Haile Selassie can be characterized as vigorous attempts at state building. Boundaries were defined and internationally recognized. The power of regional lords was gradually diminished, and centrally appointed governors were put in place. After the brief period of Italian occupation, which dealt a death blow to the armies and the prestige of the feudal lords, the Ethiopian state continued its programme of administrative penetration in the countryside. A European-type administration and a standing army were introduced and feudal obligations were transformed into direct and indirect taxes. A thin network of physical infrastructure was built.[25]

The pace of these developments was substantially increased, and their scope broadened, after the revolution in 1974. As indicated above, the state can be said to have been successful in 'capturing' the peasantry. But the efforts by the Ethiopian state to penetrate the territory and control the inhabitants have not proceeded without disturbances. The most serious challenges to the Ethiopian state are the wars in Eritrea and Tigray.

Ironically, the war in Eritrea is closely related to the most conspicuous achievement of the Ethiopian state during Haile Selassie's rule. The former Italian colony of Eritrea was federated with Ethiopia in 1952 through a United Nations resolution. Eritrea was to have its own institutions for internal affairs, but the imperial government gradually imposed its decisions on the federated territory. In 1962, the Eritrean parliament was dissolved and Eritrea was incorporated into the Ethiopian state as an administrative region ruled by a governor appointed by Addis Ababa. This created resentment and armed insurrection increased during the 1960s. During the Ethiopian revolution, it was made clear that the military government was not prepared to give Eritrea a special status, and the conflict escalated. The Eritrean People's Liberation Front (EPLF) emerged as the major organization representing the independence movement.[26]

During the last decade the EPLF has built up a military apparatus

which challenges the Ethiopian army in Eritrea, and a civil administration which operates in areas outside government control. The
EPLF claims the right to self-determination for the Eritrean people,
including secession from the Ethiopian state. Eritrea is war-torn.
Almost every year the Ethiopian armed forces mount offensives
which ravage the countryside, and which are followed by EPLF
counter-offensives.

No less serious for the Ethiopian government is the challenge
from the Tigray People's Liberation Front (TPLF). The region of
Tigray has traditionally guarded its autonomous status and resisted
the hegemony of the central government in Addis Ababa. When the
military government dismissed the traditional aristocratic lord as
regional governor in 1975 and showed signs of extending direct
control, the opposition movement in the region gained momentum.
The TPLF emerged as the major resistance movement in the late
1970s and became a military factor of importance. Since then, the
area under control of the Ethiopian army and administration has
shrunk to only a few towns and trunk roads. In 1989, the government lost its last strongholds in the region.

The TPLF also operates in the northern parts of Gondar and
Wollo provinces. Military operations roll back and forth in these
areas, devastating the countryside. Most parts of the region lack
effective administration from the central government, while the
TPLF attempts to impose its own order. The TPLF is not for
secession. The organization considers Tigray to be an integral part
of Ethiopia, but it maintains that the central government in Addis
Ababa lacks legitimacy.

Since 1989, the insurgency has expanded southward in Wollo and
spilled into Shoa. It appears that the TPLF has joined hands with
the Ethiopian People's Democratic Movement (EPDM) to form a
military front called the Ethiopian People's Revolutionary Democratic Front (EPRDF), which is taking over control of the countryside throughout Wollo and in northern Shoa.

Thus it can be argued that there is a long tradition of government
and administration in the Ethiopian highlands. During the twentieth
century the Ethiopian state built up a considerable administrative
capacity, but no consensus on the nature of the Ethiopian nation-
building project has been achieved. The authoritarian nature of the
government administration and political system precludes dialogue,
and civil strife has therefore become the only way to articulate
opposition. The government diverts the lion's share of state resources

to combat the insurgents and to reassert its authority.[27] As a result, budgetary allocations to development programmes are squeezed. In the process the development programmes have taken on an increasingly *extractive* character (including the procurement of grain at low prices, collection of taxes, unpaid labour campaigns and obligatory financial contributions to 'revolutionary' tasks) and have developed *penetrative* aspects (including villagization and resettlement schemes, collectivization of agriculture, and the establishment of state-controlled mass organizations).

The system appears to be locked in a self-defeating circuit. The state uses coercion to extract resources from the population in order to finance the military crushing of its opponents. But its authoritarian measures create more opposition. Even when drought strikes, military and security budgets are given priority in the allocation of resources. The overall result is deprivation for the people, brutalization of society and impoverishment of the country. It would be futile to think that environmental rehabilitation through people's participation is a viable alternative under such circumstances. In order to achieve lasting results in the battle against environmental degradation, the Ethiopian state would need to marshal substantial economic and human resources as well as popular support. At present the resources are not available because the priority is given to the army and state security.

TOWARDS SUSTAINABILITY?

No attempt will be made here to go into the intricacies of the concept of sustainability. Reference is made to Redclift's suggestion that sustainable development meets human needs and is capable of maintaining economic growth and conserving natural capital.[28] Instead of probing into the difficulties and complexities of the concept as such, I shall try to outline what sustainable development in the northern Ethiopian highlands would require during the 1990s in terms of political accommodation, participation of and incentives for the peasants, and modifications in the agricultural production.

Political sustainability

A political settlement between the government and the opposition movements is a precondition for large-scale environmental rehabilitation in the northern highlands. But peace alone will not solve the problem of degradation. The processes which underlie the problem remain. After a decade with two major famines, which were only separated by 'normal' food shortages and sufferings, the peasants are exhausted. It cannot be expected that the peasant society by its own inherent dynamic will propel itself into innovative development. The role of the state is hence critical in breaking the vicious circle of the degradation process. But the replacement of one central government in Addis Ababa with another, if the successors to power also have a monistic concept of the state, would not bring about much improvement. Neither would the mere cutting up of the present territory of Ethiopia into two or more new state formations.

A constructive political challenge facing Ethiopia is to find a political formula within which efficient development administration could be combined with power-sharing and cultural pluralism, not only between the educated elites of the various nationalities but also between the members of the thousands of peasant associations and the authorities at district and regional level.

Socio-economic sustainability

From a social science perspective, land degradation can be viewed as a consequence of human decision-making regarding land use.[29] Decision-making is influenced by perceived incentives and disincentives within differing time horizons. The incentive pattern is set by government policy and includes land tenure systems, administrative regulations, and pricing and marketing arrangements, among other factors. In the official Ethiopian 'socialist' context, all these regulations tend to discourage peasants from long-term investments in land rehabilitation. Moreover, since the peasantry lives under conditions of extreme poverty, the time-horizon of land use decisions is bound to be short. All efforts are concentrated on the current cropping season.

Rehabilitation is ultimately dependent on the day-to-day decision-making of peasant households. Hence the economic incentive structure and the regulations on land tenure would have to be recast so that peasants think it is worthwhile to invest their labour in land and water management projects with medium- to long-term gestation periods.

Land tenure is therefore extremely important. Legal rules related to the use, transfer and succession of land must be specified, and a machinery for enforcement established. In order to facilitate long-term sustainability, the rules must be designed to encourage household investment in permanent structures which diminish erosion, stabilize vegetation and increase water infiltration. The legal rules must be consistent with the incentive systems of taxes, prices and subsidies. The incentives must be directed to households and individuals, as well as to common resources such as village woodlots. Villagers do not plant and tend woodlots productively unless there are rules and procedures for how households can benefit from the collective enterprise by securing fuelwood and construction material for their individual needs.[30]

Ecological sustainability

In the present ecological and socio-economic conditions in the northern Ethiopian highlands, intensification of agricultural production, in the sense of greater input of labour into the conventional farming system, can neither stop further degradation nor feed the population. There is a need for innovations in the manner in which factors of production are used and combined.

The physical, biological and technical aspects of sustainable land management receive increasing attention in Ethiopia, both from the Ministry of Agriculture and from donors. It is agreed that the point of departure for improving land management practices should be the ongoing soil conservation programmes. But these programmes need to be modified and improved. Two major aspects will be mentioned here, namely vegetative soil conservation and innovative water engineering.

Soil conservation has so far emphasized physical structures such as bunds and terraces. But trials show that the efficiency of these structures can be greatly improved if they are combined with vegetative measures. Trials are underway to test means to integrate the cultivation of forage crops and shrubs with farming systems. Increased use of legumes, forage strips, alley farming and perennial herbaceous vegetation are encouraged. Peasants have responded positively to these programmes since such measures both reduce soil erosion and provide feed for livestock. This is a concrete example of a 'conservation plus production' approach.

Hillside closures also have an untapped potential for resource management. Most of the closed hillsides are left untouched, and the composition of species invading the hills is sub-optimal from a production point of view. However, the closed hillsides could become productive for fuelwood, fodder and building material. This would require the closures to be carefully managed, and divided into compartments for various purposes (such as fodder grasses, fuelwood, building poles and domestic water supplies).

Work on water conservation and irrigation is also needed. There seems to be a major potential in the micro-irrigation schemes, but overall there is a need for more innovative solutions to water engineering and year-long water storage. Techniques for *water harvesting* try to concentrate the rainwater infiltration over a wide area into a smaller area where crops are grown. Techniques for *water spreading* collect seasonal torrents in rock dams from which water is distributed to cultivated fields as supplementary irrigation.[31]

The above-mentioned options provide a broad spectrum for research that hardly fits into the conventional agricultural research system. Single factor experiments and solutions must be abandoned in favour of research combining production and conservation using a multi-factor approach. This means, *inter alia*, that research stations would need to be moved from fertile flat land to eroded slopes.

Furthermore, the development of management plans for natural resource utilization at the local level is a precondition for successful rehabilitation. This means, *inter alia*, that the 'land managers', the farmers, should be incorporated as resource persons in research programmes.

Although there still is potential for developing sustainable production systems in degraded highland areas in northern Ethiopia, the realization of this potential would require massive external support. A drastic intervention would be to take selected parts of the land out of production for some years. The exhausted state of much agricultural land in Ethiopia requires that it 'rest' and be 'doctored' before going back into production.

The consequences for the peasant society of such an intervention would be far-reaching. Not only would land have to be taken out of immediate production, but much peasant labour would have to be directed towards rehabilitation work at the expense of cultivation. This means that the food deficit would be substantial. A new approach would therefore have to be designed on the basis of long-term agreements between peasant associations, the authorities and donors. The long-term nature of rehabilitation requires a decade-long agreement, under which peasant associations would undertake to implement rehabilitation programmes in accordance with a plan of operation emanating from research findings, and agreed upon by all parties. While the peasants would leave areas out of cultivation and close them to grazing, the authorities would undertake to distribute food, in compensation for food production foregone by the substitution of rehabilitation efforts for cultivation. The food itself would have to be provided by international donors under the auspices of WFP.

This approach has the disadvantage of requiring massive external support for a period of many years. Such support is perhaps inevitable, since the peasants are now in a state of exhaustion which precludes an enthusiastic attitude toward risk-taking and innovative activities. However, it is important that the period of dependence be as short as possible.

Rehabilitation entails economic costs, and an assessment must therefore be made of its relative costs and benefits. The further degradation has proceeded, the higher the costs of rehabilitation will be, and, more importantly, the fewer the productive options will be. In hilly areas where most of the vegetation and topsoil have disappeared, the only realistic conservation measure may be the planting of drought-resistant shrubs with little production potential. The area to be rehabilitated must therefore be subdivided in accordance with differing degrees of degradation, and with varying prospects for restoring the production of crops, fibres and feed.

CONCLUDING REMARKS

These examples may suffice to indicate that there are real opportunities for ecological rehabilitation and the development of modified agricultural production systems in the northern highlands of Ethiopia. It is also evident that such sustainable development would require an active and massive effort by the millions of peasants living there. Rehabilitation can only come about through people's participation based on incentives and empowerment. But the efforts must be guided by a state administration, and resources must be brought in from external sources. Thus the future of the northern Ethiopian highlands depends on whether the people, the government and international donors can find a constructive political formula for cooperation. The outcome will spell the difference between ecological collapse and human disaster on the one hand, and gradual rehabilitation on the other.

NOTES

1 This chapter was written shortly after the recent change of government in Ethiopia.
2 This conceptualization draws on the theoretical frameworks outlined in Blaikie and Brookfield 1987, and Dixon *et al*. 1990.
3 Highlands are defined as being above 1,500 metres.
4 Westphal 1975.
5 Central Statistical Office 1985.
6 Ministry of Agriculture and FAO 1984.
7 Ibid.
8 For an exhaustive treaty on the subject see Russell 1973.
9 Ministry of Agriculture and FAO 1984; FAO/UNDP 1984.
10 Yeraswork Admassie *et al*. 1983.
11 Yeraswork Admassie 1988.
12 Office of the National Committee for Central Planning 1986; Yeraswork Admassie *et al*. 1985.
13 Yeraswork Admassie 1988. This report provides a thorough analysis of the impact of on-farm conservation structures, tree planting and hillside closures as well as a discussion of factors facilitating and hampering the social sustainability of such rehabilitation measures.
14 Fruhling 1988; Hultin 1988.
15 Markos Ezra and Kassahun Berhanu 1988.
16 Yohannes Gebre Michael 1988, quoted in Yeraswork Admassie 1988.
17 An analysis of the agricultural policy is found in Ståhl 1989.
18 In 1989 the government announced changes in agricultural policy including some agricultural liberalization of grain trade and increases in producer prices.
19 Yeraswork Admassie 1988.
20 Yeraswork Admassie 1990.
21 Pearse and Stiefel 1979.
22 Redclift 1990.
23 UNRISD 1989.
24 Teshome Kebede 1984; Pankhurst 1966.
25 Clapham 1969; Perham 1968.
26 On the emergence of liberation fronts in Eritrea, see Markakis 1988; Bondestam 1989.
27 Although no official figures are available, it is estimated that 40–50 per cent of government spending is consumed by the military apparatus and another 10 per cent by security.
28 Redclift 1987, 1990.
29 Dixon, James and Sherman 1990.
30 This point is clearly spelled out in Bandyopadhyay 1990.
31 Harrison 1987.

REFERENCES

Bandyopadhyay, J. (1990), 'From natural resource conflicts to sustainable development in the Himalaya – critical reflections on the scope for people's participation', paper presented at the UNRISD workshop on 'Sustainable development through people's participation in resource management', 9–11 May.

Blaikie, P. and Brookfield, H. (1987) *Land Degradation and Society*, London: Methuen.

Bondestam, L. (1989) *Eritrea – med rätt till självbestämmande*, Lysekil: Clavis.

Central Statistical Office (1985) *National Census*, Addis Ababa.

Clapham, C. (1969) *Haile Selassie's Government*, London: Longman.

Dixon, J.A., James, D.E. and Sherman, P.B. (1990) *The Economics of Dryland Management*, London: Earthscan.

FAO/UNDP (1984) *Ethiopia: a land resources inventory for land use planning*, Technical Reports 1–6, Rome: FAO.

Fruhling, P. (1988) *Utveckling bättre än nödhjälp. Om Röda korsets katastrofförebyggande arbete i Wollo*, Stockholm: Röda Korset.

Harrison, P. (1987) *The Greening of Africa: breaking through in the battle for land and food*, London: Paladin.

Hultin, J. (1988) *Farmers' Participation in the Wollo Programme*, Stockholm: SIDA.

Markakis, J. (1988) 'The nationalist revolution in Eritrea', *Journal of Modern African Studies* 26 (1): 51–70.

Markos Ezra and Kassahun Berhanu (1988) *A Review of the Community Forestry Programme and an Evaluation of its Achievements*, Addis Ababa: Institute of Development Research, Addis Ababa University.

Ministry of Agriculture and FAO (1984) *Ethiopian Highlands Reclamation Study*, 2 volumes, 17 working papers, Addis Ababa.

Office of the National Committee for Central Planning (1986) *Workshop on Food-for-Work in Ethiopia*, Addis Ababa.

Pankhurst, R. (1966) *State and Land in Ethiopian History*, Addis Ababa: Oxford University Press.

Pearse, A. and Stiefel, M. (1979) *Inquiry into Participation: a research approach*, Geneva: UNRISD.

Perham, M. (1968) *The Government of Ethiopia*, London: Faber & Faber.

Redclift, M. (1987) *Sustainable Development: exploring the contradictions*, London: Routledge.

Redclift, M. (1990) 'Sustainable development through popular participation: a framework for analysis', Geneva: paper presented at the UNRISD Workshop on 'Sustainable development through people's participation in resource management', 9–11 May.

Russell, E.W. (1973) *Soil Conditions and Plant Growth*, London: Longman.

Ståhl, M. (1989) 'Capturing the peasants through cooperatives: the case of Ethiopia' in B. Gyllström, and F.M. Rundquist (eds), *State, Cooperatives and Rural Change*, Lund Studies in Geography Ser. B. Human Geography 53, Lund: Lund University Press.

Teshome Kebede (1984) 'Some aspects of feudalism in Ethiopia', in Rubenson (ed.) *Proceedings of the Seventh International Conference of Ethiopian Studies*, Uppsala: Scandinavian Institute of African Studies.

UNRISD (1989) 'Sustainable development through people's participation in resource management: a research proposal', Geneva.

Westphal, E. (1975) *Agricultural Systems in Ethiopia*, Wageningen: CAPD.

Yeraswork Admassie (1988) *Impact and Sustainability of Activities for Rehabilitation of Forest, Grazing and Agricultural Lands Supported by WFP Project 2488*, Addis Ababa: report to WFP and to the Natural Resources Main Department of the Ministry of Agriculture.

Yeraswork Admassie (1990) *Emerging Paradigms of Participatory Rural Development*, Uppsala: Dept of Sociology, Uppsala University.

Yeraswork Admassie, Mulugeta Abebe, Markos Ezra and Gay, J. (1983) *Ethiopian Highlands Reclamation Study: report on the sociological survey and sociological considerations in preparing a development strategy*, Addis Ababa: Ministry of Agriculture and FAO.

Yeraswork Admassie, Soloman Gebre and Holt, J. (1985) *Food-for-Work in Ethiopia: a socio-economic survey*, Addis Ababa: Institute of Development Research, Addis Ababa University.

Yohannes Gebre Michael (1988) 'Land use, agricultural production and soil conservation methods in Andit-Tid area, Shewa region' MA thesis, Dept of Geography, Addis Ababa University, June.

12

LOCAL RESOURCE MANAGEMENT AND DEVELOPMENT: STRATEGIC DIMENSIONS OF PEOPLE'S PARTICIPATION

Philippe Egger and Jean Majeres

ENVIRONMENT AND PEOPLE'S PARTICIPATION: THE ISSUES

A number of factors, such as rapid population growth, unequal distribution of productive assets and natural calamities in the form of drought, floods, desertification and land erosion activate environmental strains and clearly aggravate poverty. Conversely, poverty alleviating programmes and policies in the form of direct intervention measures need to build on a positive interaction with the natural resource base. It is gradually being recognized that a lasting impact on poverty cannot be expected in the absence of a sustainable resource base for the poor.

People's participation is perceived today as an important dimension of an environmentally sustainable pattern of development. There are basically two reasons for this. When participation rests on some form of organization, it can encourage the direct management of local resources by the users. Secondly, such responsibility can be exercised in the collective interest embodied in the organization.

The question is whether this perception is correct, whether it can be verified, and to what extent and under what conditions participation can be a basis for an environmentally sustainable pattern of development. This study aims to document and analyse various strategic dimensions of people's participation in local resource management.

SUSTAINABLE RESOURCE MANAGEMENT: A GLOBAL VIEW

The concept of resource management is a comprehensive and complex one. It is based on the interaction of three elements, as shown in Figure 12.1, namely the physical resource base (land, water, forests, fauna, flora, the climate), the production system or the particular mix of technologies and of productive activities, and the social regulation, that is, the rules and laws which govern access to the resources, their use and distribution of benefits. The three dimensions are intimately knitted together and in constant interaction with each other.[1]

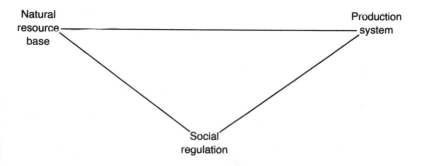

Figure 12.1   Resource management interactions

The number of input and output variables which govern the interaction between these three elements can be quite high and one can point here only to some of the more critical ones. Access to and use of resources result from a variety of property regimes; policies and tenure arrangements are determined by forms of social regulation. Types of social regulation and environmental characteristics will also influence choice of technologies and techniques which shape the production system. The interaction between the resource base and the production system will determine the observed trends in yields (for instance, of cultivated land, of forest produce, of fish catches and of pasture land). These interactions can be further defined with respect to their impact on economic productivity, their technical viability regarding renewability of the resource base, and their social implications for poverty and distributional equity.

With a growing awareness of the erosion of renewable and non-renewable resources, the orthodox economic appraisal of productive activities with regard to output, productivity and returns is gradually being challenged to incorporate non-economic criteria, such as the renewability of the resource base. The World Commission on Environment and Development has opened up new avenues by advancing the widely acclaimed principle of 'sustainable development' (WCED 1987). The term refers to a state of equilibrium between the three elements of Figure 12.1. This concept is discussed in depth elsewhere in this volume, but it will be noted here that one of its major contributions is a new examination of equilibrium and disequilibrium in resource management. Not only is the stability of the natural resource base important to consider in this context, but various social factors are crucial as well. Environmental equilibrium can be affected by the existence and strength of regulations governing access and use of resources, by the existence of local organizations which work in the interest of maintaining local livelihood, and by the choice of appropriate or inappropriate technology for resource management.

A sustainable pattern of development would therefore need to integrate growth in productivity, environmental stability and relative equity of access to productive assets and economic opportunities. In this sense, sustainable resource management could be defined as the highest productivity growth permitted by an equal distribution of economic assets and opportunities and compatible with the stability of the resource base.

Such a strategy is more likely to be developed and applied at the local level than at the national or global levels. It would require the following elements:

1 resource management which is area-specific in order to account for the characteristics of local ecosystems;
2 allowance for a full participation of the local population relying on its (traditional) forms of organization and area-specific know-how;
3 making the best use of the local resource base: human, natural, institutional, technological;
4 the adaptation of all elements of any outside intervention (rules, technology, advice) to the local situation.

This approach poses a number of dilemmas which are far from being resolved: how to develop appropriate institutional forms for local resource mobilization, how to ensure that technological choices are compatible with local management and renewability of the resource base, how to determine the economic stability and environmental soundness of development initiatives, local resource utilization and macro-regulation, and how to prioritize short-term needs, in terms of production, as opposed to the longer-term requirements of conservation.

Naturally, local strategies relate to wider processes and situations with which they interact, be these in terms of the patterns of the wider economy and its overall performance, the size and growth of the population, the forms of social and political organization, the role and policy objectives of the state or the level of scientific and technological knowledge. In the same way, people's participation in local resource management is both locally and nationally determined.

The following sections present a number of cases illustrative of positive interactions between the four constitutive elements of environmental equilibrium. The concluding section analyses several strategic dimensions of people's participation in local resource management emerging from these examples.

LOCAL RESOURCE BASED INFRASTRUCTURE DEVELOPMENT

This section discusses some examples of local resource based infrastructure development programmes implemented by the ILO. Infrastructural works are among the activities most relevant to the ecosystem: they lend themselves to planning and decision-making exercises at both the local and national levels; they involve both productive and conservationist issues and objectives; they serve the interests of specific groups of people, and thus may create conflictual situations detrimental to a coherent development process; and their efficient management typically requires people's participation.

Local development contracts in Guinea

After several decades of bureaucratic and exploitative top-down policies, the peasantry of the districts of Dabola and Dinguiraye is organizing itself in a way which (1) safeguards its autonomy in developing and managing its local resources; (2) ensures access to technical know-how; (3) allows for negotiations with public authorities and outside institutions.[2]

The approach promoted is a simple 'contract approach'. The first

307

step is for a number of peasant families to get organized into some associative forum (which may be, for instance, economic interest groups, village associations or women's groups) to carry out a joint project. The groups then enter into a written agreement with the local administration and the external organization on the basis of a detailed cost estimate, specifying the contributions of each party, its rights and obligations. The types of contracts entered into during the initial phase included: the establishment of 10-hectare plantations of grafted mango trees, orange trees and teak; swamp development; vegetable growing; school construction; latrine construction near public buildings like schools or mosques; and construction or rehabilitation of bridges, wells and other structures.[3] The common characteristics of these schemes are that they require a collective effort, and that they are of such dimensions that they can actually be managed by small groups or village associations. The village associations, through the contractual arrangement, contributed part of the costs, the rest being provided by the external organization. Technical support required and included in the contract is provided by the district administration. To make sure that the administration, which often lacks the means and incentives to work efficiently, would actually fulfil its obligations, a provision for technical support services is included in the cost estimate, under the contribution of the external organization. This provision is paid to the associations, who in turn pay the technician for the services provided.

This approach has proven highly successful during the first test phase. Local resources are developed and managed by the people themselves: human, financial, institutional resources and know-how are mobilized on a sustainable basis, in spite of the initial subsidy. The associations have all encouraged the use of local resources and materials as opposed to external or imported resources. Moreover, in order to initiate and build up the self-financing capacity of the village associations, the subsidy component is gradually reduced in any follow-up contract entered into with the same association.

The system has undoubtedly provided villages with access to know-how previously existing only at higher levels. A non-negligible by-product has been the changing relations between the village population and the government technicians, who are now literally a 'support service' to initiatives taken by the people.

In addition, this experience opens up perspectives for further development of local resource management and the introduction of wider concerns, including environmental or other long-term considerations which are normally less attractive to poor and needy populations. One of these possibilities is to extend this approach to district-level investments through the creation of a district development fund managed by a local development committee. This would enable, on the one hand, village associations to influence public decision-making and investments; on the other hand, regional or district authorities could negotiate with an organized peasantry ways to tackle larger state-wide problems such as those related to the sustainability of the resource base and the protection of the environment.

Erosion control in Rwanda

The province of Gitarama in Rwanda is experiencing extremely high population pressure in a context of limited and highly erosion-prone cultivable land. Population density is on average about 280 inhabitants per square kilometre, as compared to 180 for the country as a whole. Pressure on land is such that more than 30 per cent of the farms are of less than 0.5 of a hectare, with a substantial number below the minimum required for subsistence. All agricultural land is now occupied, and cultivation takes place on steep slopes. The result is declining land fertility and productivity, an increasingly impoverished peasantry, and – through both over-exploitation of the land and soil erosion – a steady and accelerating deterioration of the environment.

In this context, attempts have been made since the 1970s to introduce so-called 'radical terraces', radical in the sense that they are built directly as such, as opposed to the 'progressive terraces', built over several years between two anti-erosive ditches.[4] Radical terraces were first built in the provinces of Byumba and Gikongoro. They were entirely built by hand, generally 4 to 5 metres wide on slopes of 20 to 25 per cent; they present a slight declivity from the outer bank towards the inner angle; the banks, covered with grass, also have a slight slope.

This technique has proven both an efficient anti-erosion device and a means to increase land productivity. Its positive effect on soil and water conservation needs no elaboration. Run-off is completely eliminated, and several observations have shown that, in periods of drought, crops resist much better on terraced than on non-terraced land. In addition, terraces facilitate agricultural intensification: farming operations are more easily and more completely performed on horizontal land. As a result, yields have increased substantially.

However, two major constraints have so far limited the extension of the approach. First, specific technical prescriptions must be followed, above all to ensure that the top soil is removed and safeguarded during the terracing work, to be returned to the completed terrace. In addition, particular care must be taken with the construction of the bank. Therefore, peasants need some training and technical support.

Second, the labour demand of terracing is very high. Theoretically, the construction of a 5-metre wide terrace on a 20 per cent slope requires the removal of 1,020 cubic metres of soil per hectare, which under present conditions – i.e. with a labour productivity of about 1 cubic metre per work day – implies that the labour force required per hectare amounts to 1,000 work days. This effort is clearly beyond the capacity of an individual peasant.

As a consequence, ways and means must be found to sensitize the rural population, provide training and technical support, and mobilize the workforce at an acceptable cost. This is precisely what a special rural works project has set out to achieve since the mid 1980s in the province of Gitarama.

The approach currently tested consists of two stages. The first stage, highly subsidized by the project and the government, involves setting up demonstration sites in all the communes of the province. This implies an information campaign which includes visits to the experimental area in Byumba, followed by three weeks' training for a limited number of peasants per commune, who are selected from among those who have volunteered to start the works on their own land. Back home, each of the peasants trained will build 15 hectares (at the beginning this was even 0.30 of a hectare) of terraces.

The project pays for the necessary work days and tools, as well as for those of three fellow peasants who help build the terrace. These fellow peasants, in turn, are paid to carry out the same work on their own field, and to organize training sessions for interested peasants: however, this promotional work is limited to three demonstration and training sites per sector (a sub-division of the commune). The cost to the project of this subsidized promotion is about FRW 15,000 (Rwandan francs) (approximately US$180) per terrace of 0.15 of a hectare.

The second stage consists of the promotion of self-organized groups of peasants, convinced of the benefit of the terraces, and willing to carry out the terracing work through mutual help, that is, without subsidies from the project other than the tools, which are provided as an incentive.

This approach is very difficult to apply, and requires special efforts and a capacity to motivate peasants on the part of the technical staff. Yet, the results so far are positive, with more than 65 hectares of terraces built for demonstration and training purposes in 9 communes, more than 1,000 peasants trained, and more and more self-help groups being constituted, some composed only of women. Thus, with appropriate methods, long-term resource conservation practices can be introduced into individual farm management, provided peasants realize the benefits that such innovations can bring about.

Irrigation associations in Nepal

In 1988, the government of Nepal adopted an irrigation sector programme (the SPWP) whose major thrust is on farmers' participation, in recognition of the role water users need to play in sustaining irrigation development and management (Anseri 1989; Baker 1989). This approach is largely based on experience gained in a special public works programme, which started in 1980 and constructed or rehabilitated 142 small- and medium-scale hill irrigation schemes, covering some 5,800 hectares. The programme has two special characteristics, which distinguish it from ordinary public works programmes operating in Nepal: it uses labour-intensive and local resource-based technologies to build infrastructure, and it is based on popular participation. With the decision of the Nepal government to apply these principles on a wider scale, the SPWP is now concentrating on two major activities.

First, it focuses on the mobilization of local resources for small-scale irrigation schemes. Wherever feasible, these schemes are integrated into 'programme packages' and complemented by river 'training' (correction of the river bed) and embankment protection works; other types of irrigation-related rural infrastructures, such as canal line slope stabilization, overpasses, trails and water supply are considered necessary to make irrigation schemes viable. In addition, the environmental dimension, incorporating soil erosion control, afforestation and land stabilization works is now seen as an integral part of construction activities with a view to minimizing any negative ecological damage from heavy monsoon rains.

Second, the programme focuses on users' participation, involving the local population and farmers' irrigation associations in the construction, operation, management and maintenance of irrigation schemes.

During the first phase of the Nepal SPWP (1980–7) the formation in all projects of farmers' irrigation associations and their management body, the water users' committees, laid the foundation for active popular resource mobilization. The water users' committees played an important role in construction, maintenance and operation of the irrigation system.

This experience has prepared the way for the adoption of a participatory approach to irrigation development in Nepal. The approach spells out the need for the formation of a constitutional farmers' irrigation association and its Water Users' Committee, a written agreement between the association and the Department of Irrigation on responsibilities for implementing the project prior to the commencement of construction work, the placement of a minimum cash deposit at the local commercial bank, and a willingness on the part of the beneficiaries to take over the operation and maintenance of the irrigation scheme on completion.

A key element in the success of this participatory approach is the written agreement, which defines the contributions, responsibilities, rights and obligations of the farmers' irrigation association on the one side, and the Department of Irrigation on the other. This agreement is negotiated in three stages, the first of which is the request from the association for assistance. Second, following a preliminary evaluation of the project potential by the department, a proposed agreement in principle is elaborated, which commits both parties to further explore the feasibility of the project and their participation, and contains estimates of the total project costs. The final stage comes after a detailed assessment of the design, costs and time frame for completion, and the commitment of each party to incur their contribution and obligations.

Presently, six different ways of mobilizing the scarce local resources of the farming population are tested, ranging from registration fees to cash, labour or land contributions, bank loans and capitalization of previously completed construction works.

In order to implement the participatory strategy, the Department of Irrigation now needs to strengthen its ability to organize farmers into effective associations and involve them in meaningful tasks. Its role now goes beyond construction of facilities and system management, and includes building or strengthening the essential 'socio-infrastructure' of new or existing farmers' groups which could take on maximum responsibility for project construction and system operation and maintenance. This is a task that is largely new to the department, and requires significant reorientation of its technical staff, creation of new skills and formulation of appropriate procedures.

The example reveals that a constructive working relationship can be established between the peasantry and a technical community, on the basis of a mechanism (the 'contract') which provides for: *autonomous organization* of the 'target group'; *negotiation* of the responsibilities, inputs and rights of each party; *access* to technical know-how; and integration of wider concerns (such as the environment) into resource management arrangements at the local level.

PEOPLE'S PARTICIPATION IN LOCAL RESOURCE MANAGEMENT

This section examines three cases in which organized people's initiatives have led to some form of resource management. The cases are drawn from initiatives with which the ILO is in contact through its programme on participatory organizations of the rural poor.

People's development initiatives in Burkina Faso

The Yatenga region in the north west of Burkina Faso presents a typical Sahelian environment characterized by low and erratic rainfall (300–400 millimetres annually), a very degraded environment due to growing population pressure on a fragile ecosystem and the recurrence of drought. Agriculture is a risky undertaking in Yatenga and yields are very low. The region has experienced its share of development projects, which passed with time leaving little in terms of lasting effects for the rural population.

The late 1970s witnessed the emergence of two separate developments which subsequently converged. One is the birth of an aid association by the name of Six-S (an acronym from the French on how best to use the dry season in the Sahel) intent on assisting people's own development efforts (Sawadogo and Ouedraogo 1988). Six-S believes that genuine development can only start from people's own problems, priorities and resolve. It assists group initiatives with an externally funded flexible fund whose use is decided by the people and controlled by Six-S. The objective is to mobilize village people during the long dry Sahelian season, 7 to 8 months a year, for collective village development initiatives. Six-S has grounded its work in the revival of a traditional village mutual help association by the Mossi name of Naam. The Naam groups provide the local organization, managing people's initiatives and labour, whereas Six-S provides some financial support as well as the basic message that people should take development into their own hands.

The other development is an Oxfam-funded agroforestry project which aimed to test simple techniques for reducing soil erosion and water run-off (Wright and Bonkoungou 1986). One of the methods tested, again a revival of a traditional technique, consists in constructing horizontal rock bunds along the contour lines of farmers' fields. Farmers were trained in operating the simple water-tube level used to determine contour lines and mutual help groups established to carry the rocks and build the small ridges. The technique regulates water run-off and substantially improves soil and water conservation. It has a direct effect on yield increases (20 to 30 per cent) and permits the reclamation of long-degraded land.

Dissemination of the technique was facilitated by the Oxfam project with the training of a few extension workers. It was given a powerful boost when Six-S-assisted Naam groups adopted the technique and gave it all the support of their collective energy. Specific training camps were set up, in which experienced farmers teach others wanting to adopt the technique. Farmers teaching farmers has been shown to be the most powerful dissemination device. Each group which masters the technique then transmits the experience to others. Virtually all of the six hundred Naam groups in the Yatenga region have now adopted this technique. It is simple and easy to learn, and its effects are visible enough to readily convince farmers. Construction of the rock ridges is labour-intensive and requires cooperation and organization.

These are, however, not the only innovations that have been adopted by peasants through the dynamism of the Naam groups. Afforestation, vegetable gardening, access roads, small dams, livestock, literacy and health are among the many areas in which the groups are active. The Yatenga region remains an ecologically very fragile zone. However, the efforts of the Naam group are beginning to make an impact.

There are basically three lessons that can be drawn from this experience. First, environmental rehabilitation is local resource- and labour-intensive. It requires effective cooperation and work organization which only groups based on solidarity can provide. Second, sustaining a fragile ecosystem requires constant innovation and experimentation to test different techniques and approaches. Only people organized for their own development and committed to their work can ensure a working environment conducive to successful experimentation. Third, good ideas and techniques need to spread fast, and an effective system of communication and training is essential. Naam groups are enthusiastic about this work, committed to moving forward together, to exchanging ideas and experience, and to learning from each other. A constant flow of visits and encounters is proving to be the best and least costly extension system.

Fisherfolk organize for aquatic reform

There are approximately 800,000 subsistence fisherfolk in the Philippines, contributing up to half of the national fish production, yet surviving mostly on per capita incomes of less than half the poverty threshold. Most of these are working as wageworkers and shareworkers and only a minority as independent owners of some fishing equipment. The livelihood of these small fisherfolk is gravely threatened as the natural resource base for fishing, the Philippines' rich marine and inland waters, is gradually destroyed and curtailed through illegal and destructive fishing practices, privatization of fishing grounds with monopoly concession systems and fishpens, as well as competition from large-scale industrial fishing. Catches by small fisherfolk have consequently been dwindling rapidly.

One response to this dramatic impoverishment of fisherfolk and depletion of fish resources in the Philippines has been organization, particularly through a non-governmental body committed to the formation of strong self-reliant people's organizations. PROCESS is a filipino NGO formed in 1982 out of a government programme assisting participatory rural development. PROCESS collaborates with small fisherfolk, farmers and migrant sugar workers through a core group of 'facilitators'. These conduct participatory investigations encouraging the poor to document their socio-economic condition. Discussion groups are formed in the course of the people's research, and gradually organizations emerge for taking collective initiatives. Of particular importance is the legal resource development programme which seeks to raise the consciousness of the poor on their rights, the significance of law and legal procedures. PROCESS currently works with some three hundred villages in nine provinces of the Philippines (Espiritu 1986).

In Bohol province, PROCESS is assisting fisherfolk in investigating the causes and characteristics of the fishing crisis and seeking ways to improve the situation of poor fisherpeople. The rights of fisherpeople have been explored, destructive fishing practices documented and negotiations with the municipalities opened in order to reform the concessionary system excluding the poor.

A province-wide workshop was held in 1986 and was attended by a large number of fisherfolk representatives. The workshop adopted a declaration denouncing illegal and destructive fishing practices and called upon the formation and mobilization of strong people's organizations to work towards the enforcement of existing laws banning such practices. The fisherfolk organizations are also actively carrying out information campaigns to investigate new and alternative fishing technologies as well as environmental protection measures.

These and other efforts at the grassroots in various parts of the country have led to the formation of a national fisherfolk association, with some fifty thousand members campaigning for a genuine aquatic reform. Defending their right to survival, these fisherfolk cry out: 'We produce the nation's fish supply but we are dying of hunger' (Toldeo 1988). The national alliance on aquatic reform has adopted a platform seeking to put the Philippines' inland and marine waters under the control of the poor majority who depend on them for their livelihood, as well as calling for effective conservation measures to preserve and protect fishery resources.

The mobilization of the Philippines' fisherfolk continues as a new fisheries code is under discussion in parliament. Whatever new legislation is to be adopted, much will depend on the fisherfolk's own association, their awareness of their rights and their determination to have a say in their country's fishery resource management. Organization and voice represent here again essential levers towards a more sustainable and equitable management pattern of a major resource in the Philippines.

Forest resources and India's tribal people

Forest resources in India are rapidly being depleted. It has been estimated that barely 10 per cent of the country is under forest, while a minimum of 33 per cent is considered necessary for environmental stability. There are many competing demands on India's forest resources. The forest industry is supplying an expanding market for paper, construction materials and urban fuels (wood and charcoal). The livelihood of a large share of India's rural population, particularly the poor, depends on free access to biomass products such as fuel, fodder and timber collected in forests. India's 40 million tribal population (approximately 5 per cent of the total population), concentrated in the major forest belts of the country (the Himalayas, central and western belts), originally lived in complete symbiosis with the forests which house and nurture them. Still today, over 50 per cent of these people rely directly or indirectly on forests for their living, gathering food and minor forest produce, supplying cottage industries or working as wage labourers for commercial contractors (Dasgupta, 1986). The welfare of tribal people is thus highly dependent on the state of the forests and on India's forest policies. This fact is increasingly recognized now that one of the central aims of policy is to rebuild forests.

Forests were considered a resource of strategic interest during the British colonial government. As early as 1855 the movements of tribal forest dwellers were regulated. In 1894, a resolution was introduced proclaiming that forests should be used for the 'public benefit' in view of the growing timber requirements of an expanding railway network. In 1952, the Indian national forest policy classified forests into three categories: reserved, protected and private. Forest people were denied any legal rights over the forests and could only retain some traditional rights over the collection of minor forest produce.

The Forest Department has been entrusted with the responsibility of guarding the forests, whereas licences are issued to private contractors to exploit particular areas. This system has led, on the one hand, to constant conflicts between the people and forest guards, and on the other to intensive and careless over-exploitation of the forests. Forest contractors have naturally relied on forest dwellers as wage labourers for unskilled work. The position of tribals as an ethnic minority has made them vulnerable to non-compliance with minimum wage and other labour laws.

Overexploitation of forests has led to severe environmental damage and prompted the reactions of a number of popular resistance movements. Among the more recent and better known examples are the Bhoomi Sena (Maharashtra), the Chipko movement (Uttar Pradesh) and the Jharkand Mukti Morcha (Bihar). Such groups are not only protesting against the commercial exploitation of forests, they are defending their livelihood, a way of life in harmony with the forests. Tribals do not view forests as a resource that can be plundered but as a habitat to be protected and nurtured. Hence such movements are also actively engaged in afforestation programmes, and they show how forest resource management reconciles people and nature.

Today, forest policy in India is a matter of public debate. The need to undertake massive afforestation is well recognized, with ambitious social forestry programmes being carried out in some states. Ironically, tribal people are now being called upon to rebuild the forests from which they had been alienated. The forest department leases out parcels of degraded forest land to tribal households, which are entrusted with the task of rehabilitation.

The success of such efforts rests on three principles. Tribal (or non-tribal) communities must enjoy effective control over the forest and its produce. Such communities should be organized to that effect, and they should ensure an equitable distribution of the benefits for the efforts invested. The first principle is a matter of policy which is yet to be resolved. The other two principles relate to the internal organizational capacities of communities (Agarwal and Narain 1990).

Commercial exploitation of forests and tribal forest dwellers offers two radically distinct conceptions of natural resource management. The real issue is whether the rules governing the use of forest resources can strike a balance between these diverging views.

STRATEGIC DIMENSIONS OF PEOPLE'S PARTICIPATION IN LOCAL RESOURCE MANAGEMENT

The preceding examples illustrate a variety of situations in which people's participation in the management of local resources emerges as a critical dimension. The situations vary greatly from one to the other, particularly with regard to the overall setting and the characteristics of participation. Taken together, however, this diversity nevertheless points to a number of strategic dimensions regarding the interaction between people, resource management and development. Five points are discussed below.

Participatory labour-intensive environmental regeneration

For most of the world's poor, resource management is tantamount to environmental rehabilitation and protection. Many environments are fragile, overexposed and overexploited, and their rehabilitation is an urgent need. Such rehabilitation, as cases from Rwanda and Burkina Faso suggest, is a very labour-demanding undertaking. A labour-intensive approach is thereby called for, and must naturally rest on the organization and cooperation of local people. Where this organization and cooperation result from an internal drive and will of people pursuing self-defined objectives, as with Six-S in Burkina Faso, it stands on much firmer ground than when people are organized and mobilized from the outside.

An intermediary situation is illustrated by the contract approach applied in Guinea and Nepal, which facilitates the consolidation of local technical capacity and of responsible village or users' associations. Successful participatory development depends ultimately on the degree to which people have organized themselves and have had the opportunity to discuss and agree on common areas of action. The particular way in which participation is initiated, by whom and how, as well as the institutional structure within which participation unfolds, appear as critical dimensions. Investments in environmental regeneration in the absence of clearly established local responsibilities which allow for various forms of people's participation in the planning, execution, use and maintenance of such works would not yield the same results. The institutional form of such investments – particularly labour-based investments – is therefore a decisive element.

Rules of access and management of local resources

The rapid depletion of fishery resources in the Philippines is a consequence of the absence of agreed upon and enforced rules regulating fishing practices. Land erosion in the Sahel can be attributed in part to the demise of traditional land tenure systems. Sound management of forest resources in India and other countries largely rests on the observance of basic rules establishing clear responsibilities and guidelines. In fact, no natural resource can be sustained in the absence of clearly established and recognized rules determining who can do what, how and when. Illustrations of the problem of the 'commons' are numerous enough to indicate that unregulated access is tantamount to a rapid depletion of the resource base.

Rules governing access and use are generally established by national governments which have the responsibility for passing appropriate legislation. What is observed, in fact, is that such legislation is often non-existent or inadequate – land tenure in Africa is a typical example – and that enforcement can be very weak, as in India.

Poor people are by definition asset-poor, and are therefore highly dependent on public or common resources. There are a number of ways in which the poor can supplement deficient legislation. Most of these build on the one strength of the poor – the power to group together and to coalesce interests. People's organizations can press for the enforcement of whatever legislation may exist, particularly at the local level. They may adopt rules and regulations for themselves and thereby establish precedents which can inspire future legislation. People's organizations can, if sufficiently strong, press for policy reforms and changes in legislation at various levels. Finally, people's organizations can offer the opportunity to determine, through consultations and negotiations, mutual responsibilities with local and national authorities. The contract approach illustrated by the case in Guinea is one such innovative approach.

Appropriate resource management calls for local responsibility, whether in support of national legislation or for locally agreed upon rules. Such local responsibility rests more effectively in the hands of those directly concerned if and when the people are, or can be, organized to that effect.

Knowledge and experience in resource management

Resource management is an area which requires intimate knowledge of local conditions and of ecosystems, as well as experience and expertise. It is an area of holistic knowledge, pulling together various technical fields where traditional knowledge can be as relevant as modern specialized knowledge. Such integration and interdisciplinary approaches, however, are not readily forthcoming. It is difficult to recover and/or further develop traditional knowledge which may only be available from specific persons in the villages. It is not easy to encourage specialists in different technical fields to agree to work together on a common problem. It is equally difficult to get technical persons to agree to work with the people and to take account of the people's traditional knowledge.

A participatory framework can greatly facilitate such tasks. First, participatory organizations can stimulate the people to reflect on specific issues and to recover age-old knowledge and experience. The pooling of knowledge and experience is a clear strength of people's organizations. Second, a people's organization can greatly facilitate collaboration with technical specialists. Depending on the strength of the organization, the term and direction of this collaboration can be appropriately influenced by the people. This is the sense of the experience in Guinea. Finally, participatory organizations are good at sharing knowledge and experience among one another, particularly as regards dissemination of information and training in innovative and useful approaches. The widespread adoption of terracing in Rwanda and contour bunds in Burkina Faso testify to this in-built extension service among the people. It is interesting to note that participation enables good ideas to spread fast and at little cost.

Local resources and employment

The employment and incomes of the poor, particularly in rural areas, depend to a large extent on their terms of access to national resources such as land, water and forests. Rights of access and security of tenure are critical issues. Equally fundamental is the mode of resource management, that is, the rules governing resource use and the particular choice of technology. Different modes entail different consequences for the sustainability of the resource base. Rapid and destructive modes are clearly unsustainable, and imply eventual declines in employment and income. What an appropriate mode of resource use should be can be debated, and is certainly an area of current and future research. It is clear, however, that a sustainable mode of resource use, that is, one ensuring longer-term employment and income prospects, depends on three criteria:

1 broad-based access, particularly for the mass of the poor;
2 an appropriate technology mix ensuring adequate regeneration of the resource base;
3 a clearly defined and freely accepted local responsibility.

These three criteria are all 'people-intensive' in that they rely on people taking an active role as primary users of local resources. The observance of these criteria depend again on the existence of strong people's movements with the capacity to negotiate and carry out development activities. It can therefore be argued that there is a link between the sustainability of the resource base and the organization of the poor and their employment and income levels. Participatory associations will manage local resources with a longer-term perspective if and when people's livelihood is clearly dependent on a resource which they control. The complexity of such local resource management in terms of organizational capacity, decentralized decision-making, and technical know-how indicates that this can only be achieved gradually. However, the experience in Nepal with water users' committees following the labour-intensive construction of irrigation works, or that of the filipino fishing groups fighting for alternative fishing practices illustrates what is at stake.

Participatory development, local and national responsibilities

This chapter has argued in favour of clearly established local responsibilities resting with the poor in the management of local resources. This issue is not only local, however. Participatory development can only grow within a favourable – or at least neutral – political space. The ability of the poor to organize themselves freely for development purposes is determined in part by the overall political setting of each country. The poor's access to natural resources is equally dependent on national-level decisions. Many countries, particularly in Asia and Latin America, have failed to redress inequitable asset distribution. The unequal patterns of asset ownership which continue to prevail are not conducive to broad-based development nor to sound resource use. Finally, local responsibility cannot be enforced in the absence of support from the national level. Developing countries are gradually recognizing the ill-effects of over-centralization and are exploring ways to decentralize administrative structures. Local resource use stands to gain from such decentralization, although much remains to be clarified in this respect.

NOTES

1 For similar lines of reasoning, see García 1984 and Berkes 1989.
2 The concept of local resource based development grew out of the 'labour-intensive public works' programme, promoted by the ILO's World Employment Programme. The local resource based approach, while including labour-based methods, is larger in scope as it covers also local materials, equipment, institutions, know-how. Together with the concept of 'access' to resources and basic goods and services indispensable to local-level development, this approach goes back to the 'basic needs strategy' developed during the 1970s.
3 See ILO/BIT 1990.
4 See Nigg 1989.

REFERENCES

Agarwal, A. and Narain, S. (1990) *Strategies for the Involvement of Landless and Women in Afforestation: five case studies from India*, Geneva: ILO, technical corporation report.
Anseri, N.A. (1989) *ILO's Role and SPWP Principles in the Formulation of Nepal's Irrigation Sector Programme*, Kathmandu.
Baker, C.F. (1989) *ILO's Role in Establishing Farmers' Groups and Users' Participation in Nepal's Irrigation Sector Programme*, Kathmandu.
Berkes, F. (ed.) (1989) *Common Property Resources, Ecology and Community-based Sustainable Development*, London: Belhaven Press.
Dasgupta, S. (ed.) (1986) *Forest, Ecology and the Oppressed. A study from the point of view of the forest dwellers*, New Delhi: New Delhi People's Institute for Development and Training.
Espiritu, R.S. (1986) 'The unfolding of the Sarilakas experience', Geneva: ILO, mimeo.
García, R. (1984) *Food Systems and Society: a conceptual and methodological challenge*, Geneva: UNRISD.
ILO/BIT (1990) *Impact socio-économique et perspectives du programme spécial de travaux publics à haute intensité de main d'oeuvre dans les préfectures de Dabola et Dinguiraye, République de Guinée*, Geneva: BIT, June.
Nigg, U. (1989) Terminal report, 'Les terrasses radicales au Rwanda', Geneva: ILO.
Sawadogo, A.R. and Lédéa Ouedraogo, B. (1988) 'Auto-évaluation de Six S: groupements Naams dans la province du Yatenga', in Philippe Egger (ed.) *Les initiatives paysannes de développement en Afrique*, BIT, working paper 47, July.
Toldeo, R.L. (1988) 'Philippines 50,000 strong national fisherfolk organisation demands genuine aquarian reform', *IFDA Dossier*, Nyon, Switzerland, 65, May–June: 373–84.

WCED (World Commission on Environment and Development) (1987) *Our Common Future*, Oxford: Oxford University Press.

Wright, P. and Bonkoungou, E.G. (1985–6) 'Soil and water conservation as a starting point for rural forestry. The Oxfam project in Ouahigouya, Burkina Faso', *Rural Africana*, 23–4: 79–86.

13

WHO SHOULD MANAGE ENVIRONMENTAL PROBLEMS? SOME LESSONS FROM LATIN AMERICA

Charles A. Reilly

Most of the world's looming environmental threats, from groundwater contamination to climate change, are by-products of affluence. But poverty drives ecological deterioration when desperate people overexploit their resource base, sacrificing the future to salvage the present.

(Durning 1990: 144)

Environmental problems obliterate boundaries between present and future, North and South, affluent and indigent, consumers and suppliers. Can these problems become issues which convince people to change their ways and better manage natural resources? Can non-governmental organizations (NGOs) and social movements expand their stewardship and resource management skills to lead in this process? In this chapter, drawn from case studies of development projects supported by the Inter-American Foundation (IAF) in the environmental arena, I will offer a series of propositions about what we know about NGOs in general and environmental NGOs in particular, and propose an agenda for future inquiry.

MUDDLING THROUGH THE MIDDLE

In Latin America and the Caribbean, participatory natural resource management projects evolve within widely different cultural and environmental contexts. Debt, adjustment, and austerity dominate most of the economies of the region. A hesitant but widespread return to democracy is challenged by galloping hyper-inflation in many countries. Middle classes, the 'mesoi' who emerged from earlier, restricted development models, are now slipping backwards. The international demonstration effect of consumer societies eludes their grasp. In fact, many of them have become 'the new poor'. The numbers of landless agricultural labourers have multiplied thanks to export monocultures, cattle grazing, and growing concentration of land tenure. And cities reveal the mix of ills coming from both developed and undeveloped 'viruses' – air and water pollution, land speculation and migration. Sorel's *The Illusions of Progress* (1969), first published in 1908, might serve as an epitaph to these forty years of development efforts. Utopias have failed us, as have the facile paradigms of right and left, which bequeathed a legacy of debt and doubt not easily overcome. A healthy dose of pragmatism seems the best defence against current ills. Reluctantly, we are now beginning to recognize that development is reversible.

The current situation has led to a new mix of public and private roles in development activities. The state role is being redefined and diminished, although markets are still far from refined or 'perfected'. Few economies have 'gotten the prices right'. Four decades of development efforts should by now have driven home the inadequacy of either micro or macro approaches, by themselves, to contribute enduring and beneficial social change for the poor majority. Development agencies today must explore the 'meso' world which transcends the isolated micro experience without making claims to overambitious macro solutions. Neither micro 'projectitis' nor macro hubris will do – rather, it is the intermediate level that must be engaged.

For some time, the major donor agencies have presided over structural adjustment and begun to grapple with debt relief. Increasingly, calls have been made for greater involvement of the NGO community – although the expectation was that the NGO community should learn to *scale up* their activities to meet the expectations of major donors. Perhaps more fitting would be for those same donors to examine their operations, their tools of the trade, and to discover how they themselves might *scale down* to more effectively reach the poor. Micro approaches are patently insufficient and macro recipes have yielded many unintended and unwanted consequences. The rhetoric of the development community increasingly celebrates NGOs. It is time for rhetoric to be synchronized with policy, or silenced.

The intellectual enterprises which inform development action must grapple less with encapsulated community case studies or broad, inclusive theorization, to enter the daunting realm of middle range propositions about what does and what does not contribute to improved living conditions for specific human groups. A paradigm shift is required to encourage those who navigate the sea of ideas to deal with formulations and concepts floating in this middle range. Ideationally too, we must learn to muddle through the middle.

One unintended consequence of the failure of growth-dominated development is the recognition that economies, at least in Latin America, are being driven to an increasing degree by informal activities. The survival strategies of the poor have coalesced in many countries into significant aggregate contributions to their economies, and the informal sector continues to offer the only employment hope for many people. Less evident, perhaps, but no less significant, is the role of informal political activity in buttressing democracy. Highly centralized, authoritarian regimes now cede to civil society – but often the mediating social structures (labour unions, peak associations, political parties, and the like) are unavailable, discredited, unrepresentative or inadequate for the requirements of the poor in today's civil society. New actors and organizations must step into the breach.

Finally, the very concept of participation must be refined to accommodate changing boundaries and possibilities of empowerment – especially of the poor. Direct democratic participation within poor people's organizations as they grow beyond local, face-to-face communities and tackle entrenched elite interests at regional or national levels requires a shift to representative structures, and the risk that leadership and followership, policy makers and policy takers, may drift still further apart. Michel's 'iron law of oligarchy' is a persistent danger as the dictates of organizational requirements override membership preferences.

It is within the shifting web of non-governmental development organizations that the IAF has functioned for twenty years, supporting literally thousands of organizations of and for the poor which inhabit this 'middle sector'. Many of these organizations emphasize popular participation throughout the project cycle. They plan, act and reflect, and, besides offering goods and services, many function as self-guided laboratories of social experimentation. The IAF increasingly supports regional coalitions and networks, within countries and across national boundaries. Project experimentation at the regional level expresses the 'meso' orientation. Less evident, but taking shape in the background, is a quest for greater policy influence – as growing numbers of NGO leaders try their hand within the public sector – or deliberately utilize NGOs as experimental sites for activities which can be transferred to government programmes.

During the past four years, the IAF has embarked on several major reviews of the popular organizations it supports. From these studies have emerged some useful roadmaps and definitions of indigenous development NGOs. I apologize for yet another set of acronyms, but consider them useful additions to the alphabet soup of the NGO world.

PARTICIPATION VIA GSOs AND MOs

Participation has been defined as 'the organized efforts to increase control over resources and regulative institutions in given social situations on the part of groups and movements hitherto excluded from such control' (Pearse and Stiefel 1979: 8). I would like to emphasize that the resources in question can be soft as well as hard, since access by poor people to information, social networks and decision-making can equal, or even surpass in importance, their access to income and jobs. Empowerment and economic survival can come together as can development and dignity. Elements for assessing participation can be found in Thomas Carroll's study of the performance characteristics of 30 'intermediary' development organizations supported by IAF in Chile, Peru and Costa Rica. He called these grassroots support organizations or GSOs. According to his definition, a GSO is a:

> developmental entity which provides services for and channels resources to local groups of disadvantaged rural or urban households and individuals. In its capacity as an intermediary institution, a GSO provides a link between the beneficiaries and the often remote levels of government, donor and financial institutions. It may also provide services indirectly to other organizations that serve the poor or perform coordinating or networking functions.
>
> (Carroll 1991: 1–5)

Grassroots support organizations are a special breed of NGOs. The most effective ones, as Carroll points out, rank high in three areas:

1 They reach poorer segments of the population and deliver services effectively.
2 They encourage participation, responsiveness and accountability and reinforce the capacity of the bases.
3 And finally, they have wider impact as innovators, extending even to the policy arena.

Besides mapping GSOs, another series of IAF studies has been examining internal democracy of membership organizations (MOs) and social movements, to discover when and how the 'iron law of oligarchy' has been countered, and accountability introduced and sustained in representative organizations. Jonathan Fox and Luis Hernández have completed a pilot study in Mexico. Chilean, Paraguayan and Brazilian researchers will join in further participatory research on this theme. Leadership accountability and organizational autonomy require delicate balancing, as Fox and Hernández have argued cogently, through 'intermediate instances of participation' defined as 'formal or informal opportunities for members to make, carry out, and oversee important group decisions' (Fox and Hernández 1989).

Finally, a research team led by Jeffrey Avina conducted in-depth evaluations of eight production projects to determine whether and how qualitative and quantitative factors are mutually reinforcing in producing beneficial outcomes for project participants (Avina *et al.* 1989). Each of these studies deepens our understanding of the NGOs and social movements emerging in Latin American civil society in recent years.

WHAT MAKES NGOs PERFORM WELL?

Not all NGOs are created equal, nor are they equally effective in generating or distributing goods and services. Like Columbus discovering America, today the development business has discovered NGOs. Hopefully, recognition of the heterogeneity of NGOs and the quest for effective ones will follow as rapidly as did the conquistadors' search for gold. NGOs have strengths and limitations. In no way can they replace the state, but they can complement other public and private initiatives if they perform well. To identify good performers, there are a number of propositions which I have extracted from studies, evaluations and fellowship research, and from my reflection on IAF funding over time.

My observations draw principally from case studies published in monographs or in the *Grassroots Development Journal*, including studies by Jeff Avina, Tom Carroll, Jonathan Fox, Albert Hirschman, Luis Hernández, Patricia Gerez-Fernández, José Pastore and Judith Tendler on GSOs and MOs, and my own field work in Mexico, Peru, Central America and Brazil. I will defend the following propositions about the internal characteristics of the most effective GSOs and MOs the foundation has supported:

Organizational characteristics

The most effective organizations the Foundation supports:

1 concentrate on a single principal task;
2 have an empathetic, self-selected staff and strong leadership;
3 have a flexible organizational style;
4 blend social thrust with technical competence;
5 are good listeners and responsive to clients;
6 are accountable to members, i.e., rank-and-file oversee the leadership, and there is direct membership participation in group decision-making;
7 frequently enjoy influential 'sponsors' or advocates in government, finance or business circles.

Social environment

Regarding the relationship of these organizations to their social environment, the following general observations usually hold:

1 Social service GSOs reach poorer beneficiaries than production-oriented ones.
2 Social service GSOs provoke less resistance than do production and marketing oriented ones.
3 MOs and GSOs require relative autonomy: i.e., control over goal setting and programme decision-making without external domination, whether by governments, political parties, religious groups or development agencies.
4 In authoritarian regimes, MOs and GSOs shelter innovators and dissidents from unfriendly official actors.
5 In democratizing settings, MOs and GSOs furnish ideas, staff and social 'R&D' space for public agencies. They serve as sites for experimentation and schools for democracy.

In a more speculative vein, I see the following trends emerging among Latin American MOs and GSOs during the 1990s:

NGO trends

1 MOs and GSOs will tend to higher degrees of specialization, often introducing fees for services.

2 There is a trend towards federative service structures, networks and consortia which, like 'small business incubators' in the US or mid-size business networks in Italy, offer technical, managerial, educational and financial services to grassroots groups (small businesses), especially at the regional level.

3 Certain MOs have evolved to levels of technical competence and sophistication which permit them to assume tasks and functions previously managed by GSOs. The 'withering away' of the NGO/GSO (if not the state) does occasionally occur.

4 GSOs can ensure islands of experimentation before, and enclaves of quality after scaling up – when the quest for quantity by government agencies erodes quality in the delivery of goods and services.

5 MOs increasingly channel the diffuse energy of social movements and actors, articulating moderated demands and pursuing self-reliance through concrete, pragmatic activities like agricultural marketing, self-help housing, and community services in education and health.

6 As debt, structural adjustment and austerity shrink government services, organizations have less to gain from a 'diminished' state, and increasingly set aside protest and demands to concentrate on production of goods and services themselves. Protest yields to production.

7 GSOs have grown increasingly effective in negotiating pacts with governmental actors for the delivery of goods and services, especially at the local level.

These general observations on the NGO universe provide a backdrop for reflection on lessons learned from projects dealing with the environment. NGOs already play key roles in agricultural production and marketing, income generation, non-formal education, health services delivery and other approaches to poverty alleviation and empowerment of the poor. As a flexible, widespread infrastructure for grappling with problems and issues, they have strengths, actual and potential, which can contribute a great deal to environmental problem-solving. The NGO universe is indeed amorphous and, at first glance, not nearly so neat as the supposed bureaucratic, hierarchical command structure depicted on organizational charts of governmental organizations. (Gustavo Esteva (1991) of Mexico describes social movements and NGOs as 'hammocks' rather than 'networks' – the metaphor evoking flexible organizations shaped by persons and purpose rather than by geometric frameworks and organizational charts.) I make no claim that all NGOs are competent, virtuous or efficient – but many are. Let me describe some of the environmental NGOs which have appeared on the IAF radar screen during the past few years.

THE IAF ENVIRONMENTAL PORTFOLIO

The Inter-American Foundation has supported several thousand projects emanating from organizations of and for poor people who lack access to adequate income, information, social networks and participation in the decisions that affect their lives. A growing number of these action projects, more than one hundred in the past five years, grapple directly with poor people's environmental problems, whether at the primary or base group level through membership organizations (MO), among grassroots support or intermediary organizations (GSO), or through networks at the national and international level. Twenty-two per cent of IAF fellowship awards in its doctoral, masters, and Latin American and Caribbean professional programmes were for research on environmental issues in 1989 and 30 per cent in 1990 – up from less than 5 per cent two years earlier.

I am convinced that environmental problems become issues which can be better managed if the people affected by them can organize and educate themselves about attacking, if not solving these problems. But like the agricultural cycle, there is a lag time between the appearance of environmental problems, identification of issues, and mobilization for problem-solving at local, regional and national levels. External intervention can help or hinder the 'appropriation' of the process by the people most affected by it. Hence the importance of examining carefully the roles of 'insiders' and 'outsiders' in this process.

Environmental projects are wide-ranging, reflecting both the gravity of problems and the creativity of the NGO sector. The projects include environmental education, agroecology, social forestry, programmes on coastal ecology among artisanal fisherfolk and seaweed gatherers, watershed management, environmental legislation efforts, environmental protection and conservation, remnants of an earlier generation of appropriate technologies, energy conservation and recycling efforts. All the primitive elements (fire, water, earth and air) are included as well as a host of modern 'viruses', unwelcome by-products of development and modernization like industrial pollution or liquid, solid and even toxic waste.

The following typology illustrates the range of projects supported by the IAF in this realm.

- *Agroecology*: crop diversification, control of agrotoxins and pesticides, cover crops, genetic diversity, organic gardening
- *Aquaculture*: maritime ecology, artisanal fishing techniques, algae, fish culture
- *'Classical' conservation*: iguana, turtles, tourism, ecological parks
- *Environmental education*: publications, videos, conferences, training
- *Forestry*: reforestation, community nurseries, extractive reserves
- *Soil conservation*: recuperation, management
- *Water*: watershed management, recycling
- *Waste management*: garbage collection, sorting, recycling

There are many cultures within Latin America and the Caribbean, considerable diversity within most countries, and a host of distinct ecological settings throughout the region. The more than a hundred IAF 'ecodevelopment' projects analysed here are spread through twenty-two countries. This distribution reflects the vitality of environmental NGOs and their networks, as well as the ability of IAF field representatives to find and fund viable projects. While the number of environmental projects supported has grown to 109 in the past five years, such projects still represent a relatively low percentage of IAF projects funded during this period. A number of people at the IAF, in particular, a working group on 'sustainable agriculture', will be further examining environmental projects, focusing on one particular cluster where experience has accumulated: organic or regenerative agriculture. The sustainable agriculture group has been examining thirty-three projects in agricultural production projects which avoid entirely or seek to greatly reduce the use of agrochemicals.

Some partial explanations for a low aggregate profile in environmental projects include:

- the responsive posture of IAF vis-a-vis proposals (the IAF does not impose but responds to project proposals initiated by indigenous NGOs);
- the IAF's institutional aversion to fads and fashions – especially those generated in the North;
- the disappointing performance of many 'appropriate technology' projects of the 1970s and 1980s – especially those created during the energy crises;
- the lag time between environmental problem recognition and its conversion into problem-solving projects or policy issues;
- technical dimensions of environmental problems which often require specializations lacking in grassroots groups.

The last point merits greater reflection. The IAF environmental portfolio has a somewhat greater proportion of GSOs to MOs than does the total portfolio. Attacking escalating environmental problems often requires expertise beyond what is readily available within membership organizations. The available store of indigenous technical knowledge may not always measure up to new levels of environmental difficulties. The importance of ideas, information and expertise in grappling with environmental threats is confirmed by the growing number of requests for fellowship support.

The issue is not simply information, but how that information gets shared. In her review of forty-four 'ecodevelopment' projects, Denise Stanley (1990) noted ten which were particularly effective, with high rates of participation by the people involved. They performed well because 'they introduced improved agricultural-forest practices to increase farmer incomes', they were 'appropriate to small farmers' land, capital and labour', and finally, 'nearly all these projects included the farmer-first approach to extension and information sharing'. Robert Chambers's (Chambers et al. 1989) advocacy of 'farmer-first' approaches to innovation and technology transfer receives strong endorsement in this cluster of projects.

Western agriculture and the green revolution brought a mix of blessings and curses to the region. Are there viable alternatives? Do ancient agricultural approaches hold secrets for contemporary needs? This 'restorationist' thrust of some organizations in Latin America has also met with mixed results. In Bolivia, raised bed agriculture as practised by the ancient Incas has withstood the test of time and continues to produce on a small scale, celebrated in articles and a public television programme (Straughan 1991). Efforts to transfer Aztec chinampa technology to the humid lowlands of Mexico did not prove so successful (Chapin 1988). As the IAF's environment portfolio expands, and it seeks constructive ties to Southern and international environmentalists, it is also among the first US government agencies to recycle paper and to purchase recycled paper for its considerable routine requirements. Environmental concern, like charity, begins at home.

ILLUSTRATIVE CASES

Let me invite you to a rapid tour of the region for a glimpse into local efforts to rescue the environment. The following descriptions of projects illustrate approaches to participatory natural resource management I have observed or which are described by colleagues or researchers. From these case studies, and other comparative research projects, I will extract a series of propositions about NGOs, in particular those concerned with environmental issues.

1 Brazil's babassu breakers: from human rights to extractive reserves

For thousands of women in Maranhão, a vast state where Brazil's lush Amazon meets the semi-arid north east, splitting babassu coconuts to extract the nut is a way of life and livelihood. A member of the palm family, the amazing babassu has myriad uses. Its leaves are used for roofing, the shell of its coconut for making charcoal, and its nut for producing a high-quality oil. Fifty-three per cent of Brazil's babassu forests are found in Maranhão, and 75 per cent of the market share in babassu nuts comes from that state.

It is estimated that close to 400,000 families in Maranhão depend upon babassu for their survival. As small or tenant farmers, they grow subsistence crops to feed their families. Yet it is babassu that provides crucial cash income to buy the necessities of life. Prices for babassu oil are low and even lower for the raw nut and charcoal, but it is one of the few resources that the rural poor can depend on. Moreover, since the collection and processing of the babassu nut is almost exclusively the province of women, it serves as a fundamental source of female employment.

In 1985, the Sociedade Maranhense de Defensa dos Direitos

Humanos (SMDDH) joined a nationwide network of non-governmental organizations dedicated to providing poor rural farmers with information about low-cost, alternative technologies. Since then, the SMDDH has offered state-wide courses on agricultural techniques, conducted a weekly radio education programme, researched and tested new farming methods and crops, and coordinated a network of 17 local organizations working in agricultural extension. Recognizing the importance of babassu to Maranhão's rural population, the SMDDH organized a state-wide meeting of the babassu breakers and translated their concerns into a project proposal for the IAF.

The SMDDH is a grassroots support organization (GSO) working with twenty-five groups of breakers in order to introduce improvements in the way the nuts are traditionally extracted and processed. At the same time, the groups are being strengthened in organization, legal matters, and administration through on-site training and a series of regional and state-level seminars. Some groups may merge with local rural workers' unions while others will seek official status as producers' associations or cooperatives. The primary concerns of these groups are to obtain better prices for babassu and its by-products and to strive for *coco livre* – freedom to collect babassu without being accused of trespassing.

Babassu is a truly renewable resource. Its primary uses do not require felling the tree, and it replants itself. Brazil is increasingly looking towards such 'extractive' resources as it strives to find ways to derive economic benefit from tropical forests while minimizing deforestation. Recognizing that meeting this challenge starts at the grassroots, the SMDDH will stimulate exchanges between the babassu breakers and producers working with rubber, Brazil nuts, and other extractive resources (Smith 1990).

2 Colombian watershed management: many partners

Dwindling water supplies have forced development institutions in the department of El Quindio, Colombia, to raise their sights – raise them above 6,000 feet – well beyond the scope of any one agency. Diminished water flow in the Quindio River, which gives its name to this small coffee-growing department in Colombia's central cordillera concerns the municipal authorities, thousands of small coffee growers who are the backbone of the area's economy, and the nation's conservation groups, who see the dwindling water flow as resulting from serious damage to the environment.

The central cordillera is the most severely deforested and eroded of Colombia's mountain ranges. Clear-cutting of forests for pasture land and overgrazing have destroyed the topsoil on the rugged hillsides that form the central watershed defined by the boundaries of the department of El Quindio.

Until recently, little heed was paid to the upper reaches of the cordillera, where only peasant homesteaders and cattle ranchers attempted to make a living. But as both municipal agencies and business leaders struggle to meet their clients' needs for water over the next two decades, attention has been focused on the region above 6,000 feet, where remaining cloud forests and stands of native trees capture vital precipitation. The future of the watershed depends on halting erosion, deforestation and pollution, and restoring tree cover and vegetation to the steep hillsides. This undertaking requires collaboration between the public and private sectors, and a deft combination of technical skills and cultural sensitivity.

The catalyst for a pilot programme was provided by Fundación Natura, founded in 1984 by Colombian businesspeople and civic leaders to conserve and manage Colombia's natural resources. An associate of the Nature Conservancy, a US conservation organization, Fundación Natura is a GSO which supports biological diversity, but also recognizes the need to protect the livelihoods of those who depend on forests and wildlands. The regional development authorities of Quindio contacted the organization to assist their budding effort to reverse the deterioration of Quindio's watershed. As a result, an autonomous watershed commission was formed to channel the resources and the technical skills of the department's Regional Development Corporation and the private Coffee Growers' Federation, and the environmental and resource management expertise of Natura's staff.

The success of this project depends on the acceptance of the rehabilitation plan by local *campesinos*, who have settled the upper reaches of the watershed. Any recommended changes in agricultural practices must take into account the immediate need of the homesteaders, as well as their long-term interests. Fundación Natura supports the community outreach and education efforts crucial to the programme. Natura will also work with local agencies and grassroots organizations to encourage innovations, such as intercropping and agroforestry that maintain production without destroying fragile topsoil. While the damage is already considerable, it is not beyond remedy. The Quindio experiment is a bellwether case for Colombia, one facilitated by excellent baseline data, the self-contained nature of the Quindio watershed, the expertise of Fundación Natura, and the innovative synthesis of public and private sector investment (Ritchey-Vance 1988).

3 Colombia: 'turning community waste to profit'

In 1979, residents of Rafael Núñez barrio, in the south-western zone of Cartegena, a city on Colombia's Caribbean coast, decided that something had to be done about intolerable living conditions in their neighbourhoods. Located on swampland that drains run off from the city, the *barrio* had been settled by squatter families who filled in patches of land upon which they built shacks of wooden stakes and cardboard. Proximity to the sea and seasonally heavy rains caused recurrent flooding, however, turning whole blocks into cesspools. A few neighbours banded together to form a membership organization (MO) called Fundación Renacer, and using sand and materials provided by a public agency, organized work teams to fill in high ground for streets.

In the following years, the Fundación held raffles and dances to raise money for a potable water system covering a five-block area. The dynamic group has also set up a revolving fund and started a concrete block factory to facilitate self-help housing, filled in land to build a day care facility for three hundred children, established a health post for children and pregnant women, and sponsored a programme to promote family gardens. With an IAF grant, the fundación will now pioneer a sanitation business that promises to make the *barrio* a healthier place to live while generating much-needed income for other community services.

Thus far, efforts to improve household sanitation have been patchwork solutions, at best. Some families have installed latrines or septic tanks in the small patios of their backyards, but most are not enclosed and frequently overflow during heavy rains. Even enclosed facilities eventually fill up, and holes for new ones must be dug. When there is no more space, families are forced to use the nineteen open drainage pipes that cross the *barrio*. When these clog, the neighbourhood is in danger of drowning in its own refuse.

To make long-term household sanitation feasible, the Fundación will buy a truck equipped with small holding tanks, a small pump, and a hose, and will open a septic tank cleaning service. Beginning with a household survey, an intensive public education campaign parallels the new enterprise. Families who already have septic tanks are referred to the new business for cleaning services. Project promoters explain the need for tanks to those families who lack them, and help organize work teams to keep streets swept and the public drainage cleaned. A variety of forums – from friends gathering in a neighbour's home, to community meetings, to street theatre performances – are used to communicate the need for personal hygiene, the importance of public sanitation in reducing illness among children, and techniques for constructing and maintaining enclosed septic tanks.

In addition to the income generated locally from its cleaning services, the Fundacion also anticipates developing new markets. The south-eastern zone of Cartegena alone contains some eleven thousand families with the same kinds of sewage problems and the same need for service as the people of Rafael Núñez *barrio*. And there are also plans to explore the possibility of turning the waste itself into an asset. After a period of time, sludge pumped from the bottom of the hermetically sealed septic tanks can be easily and safely processed into competitively priced fertilizer for marketing to nurseries and commercial farmers outside of the *barrio*.

Profits from the new business will be used to capitalize the Fundación's revolving fund to build more septic tanks, to buy medicine for its health post, and to provide scholarships to its preschool programme. The health post and day care centre have already served as models for efforts by community groups in neighbouring barrios, and success is contagious. The Fundación Renacer may be inventing a way for other groups to clean up their communities too (Ritchey-Vance 1989).

4   Spreading agroecology through research, gardens and networks

A little over three years ago, at the invitation of the Centro de Educación y Tecnología (CET) and other non-governmental organizations in Chile, a group of professionals and university professors formed the Comisión de Investigación en Agricultura Alternativa (CIAL). The CIAL group was to establish a research programme

– primarily in organic agriculture – that would lead to techniques appropriate for small farmers and people of limited resources. CET is a GSO which offers training in organic agriculture techniques to small farmers and urban gardeners at several demonstration farms in Chile.

Research at the universities and CET-run model farms have led to various production alternatives for the small landholder, and even for urban dwellers. One example is the refinement of the intensive family garden concept, in which large quantities of vegetables can be grown in very small spaces. Another is the organic subsistence farm, which demonstrates the possibility of agricultural self-sufficiency on plots of just one-half hectare.

To assure that its research will be applied at the base level, CIAL coordinates closely with non-governmental development organizations and responds to the issues addressed through their programmes. As a result, four major areas of research have been identified as posing the greatest problem for small farmers: crop production, livestock production, protection of crops, and soil management. The CET–CIAL collaborative venture has evolved into a hub for a Latin American network (CLADES: Consorcio Latino Americano sobre Agroecolgía y Desarrollo) of more than twenty NGOs involved in regenerative agriculture from eight countries. IAF support for this network of organizations promoting training and research in agroecological approaches marks a new phase in IAF's institutional evolution – more proactive in contributing to the evolution of ideas and innovations throughout the hemisphere (Page 1986: 42–3).

5 From green belt to environmental legislation

Lurin Valley, Peru – can a green belt near Lima be made productive by peasants (and can they resist buy-outs by city folk)? In Peru, fitful attempts at agrarian reform during the 1960s have fallen into disarray. A series of well-meaning but inefficient public agencies failed to resolve conflicts between workers and managers in state-created cooperatives, and many of them went bankrupt. Nearly 440 displaced workers, members of one such agrarian reform cooperative in the Lurin Valley, 30 kilometres south of Lima, sought help from a tough-minded lawyer to help them sub-divide the land and convert their cooperative to individual titles. They agreed to subdivide the land into family sized plots of 3–4 hectares. In 1983, the lawyers formed a GSO called Proterra and began an experimental 'integrated development project' which aimed:

- to secure land tenure for *parceleros* by gaining legal title;
- to improve small farmer productivity and incomes;
- to meld developmental and environmental concerns.

Land titling and related legal and administrative assistance was the first task, followed by agricultural extension activities. Crop productivity was very low and crop selection poor. Few of the beneficiaries had experience of running their own farms. Gradually, through demonstration plots and training courses, the members began to diversify, complementing subsistence crops with fruit and nut trees. Proterra technicians were convinced that more intensive production of fruit and nuts would make possible the conversion of 3 and 4 hectare plots into relatively viable economic units for families. But the legal and agricultural production challenges were only the beginning.

It quickly became clear that environmental issues were paramount. The Pacific desert which surrounds Lima makes river valleys such as the Lurin one of the few remaining green belts near the city. Like Peru's thirty-two other river basins, it is a fragile ecological setting, vulnerable to mining wastes dumped in the Andean headwaters, agricultural residue, industrial waste or urban crowding along the river's course.

Once the Proterra lawyers had proven their mettle in the land tenure arena, they became directly involved in proposing and advocating a series of new laws regarding land tenure and ecological preservation. One such law led to the creation of a green belt around the Lima metropolitan area to protect remaining farmland from urbanization. The foray into the legislative arena prompted the lawyers to do further research on conservation-related issues. Proterra has published a book on Peruvian conservation law, has compiled a directory of public and private agencies concerned with Peruvian environmental issues, and has organized a number of conferences on the environment (Avina *et al.* 1989; Carroll 1991).

ENVIRONMENTAL PROPOSITIONS

A human rights organization advocating extractive reserves for babassu coconut gatherers in Brazil, multiclass coalitions preserving a Colombian watershed from erosion, or neighbours in Cartagena mobilizing to clean up their city and trying to turn a profit from garbage – these are a few effective NGO efforts. Chilean researchers show small farmers and urban gardeners how to coax produce from land that has been spared petrochemical baths; even lawyers learn to prove their worth by helping Peruvian peasants obtain land titles, shift to more valuable crops, and then write legislation buttressing their argument that sound environmental practice and agricultural production can mesh.

What can we summarize from this scattering of Latin American cases, as well as others described elsewhere in this volume? The following are some preliminary environmental propositions and questions I would like to put forward, based on these recent case studies and my own participant observation.

Proposition 1

As democratization proceeds haltingly in Latin America, much of the social energy and normative concern formerly invested in human rights has been transferred to the environmental arena. A sound and sustainable environment is indeed a human right. Brazil's Maranhão Society for Human Rights, cited above, exemplifies well the transition from human rights struggle to legitimating alternative forms of productive organization like extractive reserves (Smith 1990).

Proposition 2

Rather than impose exogenous models, donors should look first to indigenous social organizations as development vehicles, without romanticizing them. While indigenous technical knowledge and social organization may have been well synchronized with the environment at one time, changing demographics, minifundization, insurrection and generational shifts among indigenous peoples, have often disrupted previous balances (Breslin and Chapin 1984; Chapin 1988; Straughan 1991).

Proposition 3

There are no hermetically sealed compartments separating emergency relief from development. Natural disasters often occasion effective social mobilization that can be harnessed for the development long haul. To use a phrase by Albert O. Hirschman (1984), 'aggression by nature' often triggers the emergence of cooperative action.

Proposition 4

Multiclass alliances among the organized poor and middle-class populations alarmed over environmental degradation makes for more effective empowerment. In Durango, Mexico, a 20-year process of popular sector mobilization, land invasion and confrontation has been channelled into a strong alliance between organized squatters, middle classes and local government officials confronting the causes of industrial and urban wastes (Moguel 1989; Moguel and Velázquez this volume).

Proposition 5

Environmental initiatives by grassroots support organizations venturing into the legislative arena usually require 'friends in high places', technical expertise, as well as demonstrated talent for working among popular groups if they are to be heard. Proterra, the Peruvian environmental organization did not introduce its landmark environmental legislation until it had gained credibility by introducing fruit and nut trees for farmers of small plots (Avina et al. 1989; Carroll 1991).

Proposition 6

MOs and GSOs serve as experimental sites for public programme initiatives, and have a proven ability to effectively disseminate successful experiences at low cost through their own networks. GSOs have proven themselves as effective extension services. Innovations in agroecological experimentation and extension circulate effectively through NGO networks linking research and activist organizations (Page 1986).

Proposition 7

International environmental organizations and movements contribute most effectively within the region when they work in association with indigenous or national NGOs (Carroll and Baitenmann 1987; Ritchey-Vance 1988).

Proposition 8

Waste disposal plagues Third World settings as it does the First World. Waste management is big business in the North, and solid, liquid and even toxic wastes may afford poor people's organizations a chance to make money while cleaning up the environment (Ritchey-Vance 1988).

FROM ANSWERS TO QUESTIONS

These propositions emerge from reflection on specific NGO projects in the IAF portfolio. They provide some answers and provoke further dilemmas. Some questions for systematic inquiry which still lie before us, especially at the level of comparative generalization, include the following:

Query 1

What are the salient differences between resource (mis)management emanating from above and that preceding from below? Why do we blame victims or equate the peccadillos of litterers with the disasters of industrial pollution?

Query 2

How can the interests of the local territorial community be dovetailed with the commonweal of the larger community, and who decides? What are examples of effective bargaining and fair negotiation between local organizations and more powerful institutions and interests? Which schemes for constructing dams, national parks and/or biosphere reserves, have successfully and fairly resolved the displacement issue of local populations? What new mediating structures are required in such situations?

Query 3

When might the urgency of ecological or environmental issues justify 'outside' intervention, disregarding communication and pedagogies to obtain understanding and commitment by the intended beneficiaries? On the other hand, does a romantic approach to popular participation by activists impede education on environmental issues?

Query 4

How can problems and issues become framed so as to marshall positive joint action rather than resistance and reaction by national and local elites? We are witnesses to vanishing boundaries and redefinition of sovereignty in the global village on environmental questions. How can joint action be mobilized without stirring up negative reactions across class and national boundaries?

Query 5

What are the gender-specific dimensions of understanding and organizing for environmental problem-solving? It is usually women who carry water and gather firewood in rural zones, women who have planted gardens in the slums and nurtured seedlings in the fields. And it is women who first confront the health problems of environmental degradation. Why the neglect of gender in research agendas?

Query 6

What imaginative solutions are at hand for the rural (and urban) landless – whose skills and experience are tied to a (for them) disappearing resource? What opportunities remain for off-farm employment? Should subsistence agriculture be so casually dismissed from development schemes? Do the dictates of international markets and comparative advantage guarantee expulsion of people from the land to cities unable to absorb unskilled labour?

Query 7

Is there a better way to gather information about the effective transfer of technology from scientific research centres to demonstration plots to peasants' fields? Can the breakthroughs of genetic engineering find application among small farmers as well as international conglomerates?

CONCLUSIONS

While projects tackling environmental issues raise new issues and problems, in some respects they do not radically depart from other problem-solving arenas (health, marketing, micro-enterprise promotion, etc.) and usually cross-cut them. Poor people know resources are scarce. They organize to husband resources and their organizations themselves become resources. More than certain other arenas, environmental projects require a constant flow of scientific information which does require technical expertise. If they are to work well, these organizations must find a way to promote broad-based participation and to digest the necessary flow of technical information. Where old solutions are at hand, let them be tried to see if they withstand the tests of contemporary settings. Where new problems predominate, new solutions will have to be found, tried, and effectively communicated.

Environmental problems abound. Too often, they are defined by persons disconnected from the development process. Problem-solving and issue resolution require more enlightened and direct exchange between 'insiders' and 'outsiders'. Resource management and popular participation should be inseparable. Organizations of and for the poor are the obvious, available vehicle for keeping them together.

REFERENCES

Avina, Jeffrey M., *et al.* (1989) *Evaluating the Impact of the Inter-American Foundation: case studies of eight projects*, Rosslyn, Va.: IAF.

Breslin, Patrick and Chapin, Mac (1984) 'Conservation Kuna Style', *Grassroots Development Journal* 8(2): 26–35.

Carroll, Thomas (1991) *Tending the Grassroots: performance of intermediary non-governmental organizations*, Rosslyn, Va.: IAF.

Carroll, Thomas and Baitenmann, Helga (1987) 'Organizing through technology: a case from Costa Rica', *Grassroots Development Journal*, 11 (2): 12.

Chambers, R. *et al.* (1989) *Farmer First: farmer innovation and agricultural research*, London: Intermediate Technology Development Group.

Chapin, Mac (1988) 'The seduction of models', *Grassroots Development Journal*, 12(1): 8–17.

Durning, Alan (1990) *State of the World, 1990*, New York: Norton.

Esteva, Gustavo (1991) *Comments at International Development Conference*, Washington, DC, January.

Fox, Jonathan, and Hernández, Luis (1989) 'Offsetting the iron law of oligarchy', *Grassroots Development Journal*, 13(2): 8.

Gerez, Fernández Patricia (1990) 'Movimientos y luchas ecologistas en México', mimeo, Rosslyn, Va.: IAF.

Hirschman, Albert O. (1983) 'The principle of conservation and mutation of social energy', *Grassroots Development Journal*, 7(2): 2.

Hirschman, Albert O. (1984) *Getting Ahead Collectively: grassroots experiences in Latin America*, New York: Pergamon Press.

Moguel, Julio (1989) 'Organización social y alternativos de desarrollo desde la base en Durango', mimeo, Rosslyn, Va.: IAF.

Page, Diana (1986) 'Growing hope in Santiago's gardens', *Grassroots Development Journal*, 10(2): 38–44.

Pastore, José (1988) *A Fundação Interamericana No Brasil*, Rosslyn, Va.: IAF.

Pearse, Andrew and Stiefel, Matthias, (1979) *Inquiry into Participation: a research approach*, Geneva: UNRISD.

Reilly, Charles (1985) 'Who learns what, when, how?' in William Derman and Scott Whiteford (eds) in *Social Impact Analysis and Economic Development Planning in the Third World*, Boulder: Westview Press.

Ritchey-Vance, Marion (1988) 'Turning community waste to profit', *IAF Annual Report 1987*, 45–6, Rosslyn, Va.: IAF.

Ritchey-Vance, Marion (1989) 'Helping development mesh – not clash – with conservation', *IAF Annual Report 1988*, 49–50, Rosslyn, Va.: IAF.

Smith, Brad (1990) 'Reducing the risks and increasing the gains for Brazil's babacu breakers', *IAF Annual Report 1989*, 23–4, Rosslyn, Va.: IAF.

Sorel, Georges (1969) *The Illusion of Progress*, Berkeley: University of California Press.

Stanley, Denise L. (1990) 'Environmental technology adoption and sustainability: an analysis of Inter-American Foundation projects', mimeo, Rosslyn, Va.: IAF.

Straughan, Baird (1991) 'The secrets of ancient Tiwanaku are benefiting today's Bolivia', *Smithsonian Magazine*, February: 38–45.

Tendler, Judith (1981) *Rural Credit and Foundation Style*, Rosslyn, Va.: IAF.

INDEX